BONEHEADS & BRAINIACS

Heroes and Scoundrels of the Nobel Prize in Medicine

Moira Dolan, MD

Fresno, California

Boneheads and Brainiacs: Heroes and Scoundrels of the Nobel Prize in Medicine

Published by Quill Driver Books
An imprint of Linden Publishing
2006 South Mary Street, Fresno, California 93721
(559) 233-6633 / (800) 345-4447
QuillDriverBooks.com

Quill Driver Books and Colophon are trademarks of
Linden Publishing, Inc.

ISBN 978-1-61035-350-2

135798642

Printed in the United States of America
on acid-free paper.

Library of Congress Cataloging-in-Publication Data on file.

Contents

Preface

This book covers the first fifty years of the Nobel Prize in Physiology or Medicine—from 1901 to 1950—a period that saw two world wars, the atomic bomb, and Nazism. These horrific events gave plenty of opportunity for boneheads to shine and brainiacs to struggle. On hearing "the Nobel Prize," most people think of a prestigious award for the loftiest achievements in the world given to the best of the best who have made the most useful contributions to the betterment of humankind. It is natural to idolize our heroes, so there is a tendency to presume that prizewinners are the smartest, kindest, fairest, and overall preeminent people in the world. The Nobel Prize conjures images of selfless geniuses dedicating their entire lives to a relentless pursuit of advancing knowledge for the benefit of all the peoples on Earth.

These were certainly my presumptions when I first heard about the Nobel Prize in Medicine. I was a teenager eagerly applying myself to my studies with the goal of becoming a doctor. My role model was Albert Schweitzer, a French-German physician and theologian who made his career in the deepest reaches of Africa, bringing hope and healing as a Christian medical missionary. Schweitzer won the 1952 Nobel Peace Prize for his philosophy of "reverence for life" in which he bucked the modern trend of science that had sunk into materialism devoid of ethical responsibility. Schweitzer lived his creed by placing human rights and dignity above all else.

It was a shock then to be faced with cynicism and contempt when it came time for my medical school admission interviews. I was repeatedly asked, "Why do you want to be a doctor?"

"To heal people, of course!" I readily replied.

When this drew snickers and once even derisive laughter, I gradually changed my answer to just generally wanting to help people. That only caused cynical head shaking, but a few schools accepted me nonetheless.

As a first-year med student, I spent a great deal of time in the anatomy lab performing dissection. I was good at it and deftly went about our daily assignments in a workmanlike fashion, cutting through the gut, the heart, the brain. This attitude carried me right up until the last week, when we had to dissect the feet. Just as I was about to cut into the sole of the foot, I noticed that my cadaver had a Dr. Scholl's corn plaster on his little toe. It made me consider the life of the person who had been, and I was infused with a sense of the "soul" of the man. I lost all enthusiasm for hacking into his body because it suddenly felt like an irreverent and callous activity.

Another memorable event occurred in my second year, when our class was treated to a visit from sales representatives of the drugmaker Eli Lilly. They distributed expensive stethoscopes to the entire student body, totally free. Free in the short run, yes, but it was Lilly's attempt to begin forging the bonds that would influence prescribing habits for a professional lifetime. When I questioned the ethics of accepting these gifts, I was in the definite minority. Little by little, my lofty image of the noble calling of medicine was being eroded.

By my third year, well into clinical hands-on medicine with real live patients, I noticed gross insensitivity had worked its way into our everyday language on the medical wards. "The juvenile diabetic" we called the twelve-year-old girl admitted with seizures due to her sugar being out of whack. The seventy-five-year-old man transferred from the nursing home with a gangrene infection was simply referred to as "the black foot on 4North." Then there were "the heart attacks," "the fevers of unknown origin," and, of course, always a dreadful assortment of "nutcases."

I was in my internship and residency program in the years before the law mandated that patients be asked at the time of hospital admission for their advanced directives, or living wills. Therefore, every patient who stopped breathing was immediately subjected to a resuscitation effort, complete with chest compressions and attachment to machines for artificial respiration. Most of these efforts failed, but occasionally there would be a survivor who was stuck on a machine with so much loss of blood flow to the brain that he or she would never recover a meaningful life. The general attitude was to take no responsibility for such outcomes: hey, we did our jobs. My experiences of modern medical care were drifting further away from my ideals, which seemed increasingly unrealistic.

One day we were engaged in a resuscitation effort on a ninety-five-year-old woman with "multi-organ failure"—she was simply dying. The attending physician phoned in and insisted we cease and desist. He then told me that, decades ago, the patient had been on a team of Nobel Prize–winning researchers who discovered vitamin B_{12}, and if she survived our CPR, she would surely be brain dead. What kind of life is that for such a stellar mind? It rekindled my interest in Nobel Prize winners in the field of medicine. By then I was disillusioned by the insensitivity of modern medical care but still held on to ideas that once upon a time medicine was widely practiced as a noble profession. I returned to the writings of Albert Schweitzer, who had this to say about being compassionate: "The purpose of human life is to serve and to show compassion and the will to help others." Schweitzer warned, "If a man loses his reverence for any part of life, he will lose his reverence for all of life."[1]

After thirty years in medicine, I now have the luxury to write about my fascination with the Nobel Prize. I discovered some interesting histories of the prize-winning characters little known outside the academic world, based on a great deal of contemporary and autobiographical accounts. This is thanks to both the resources available through out-of-print book finds on the internet and the policy of the Nobel committee to unlock sealed records after fifty years have elapsed.

I was alternately delighted, surprised, and dismayed by what I found. There were plenty of noble characters among Nobel Prize winners, but alas I discovered that some of my historical medical heroes were far from the honorable stars I had idealized. What I learned along the way has changed my view of the Nobel Prize in that I no longer assume the winners are necessarily the best or the most honorable or the fairest in the land, but it hasn't dampened my keen interest in their fascinating stories. I hope you enjoy reading this as much as I enjoyed the process of composing it.

Moira Dolan, MD
Austin, Texas

1. Harold E. Robles, comp., *Reverence for Life: The Words of Albert Schweitzer* (San Francisco: HarperCollins, 1993).

Introduction

The Nobel Prize

Cartridge filled with dynamite, with Nobel's signature on the paper wrapping

The whole of my remaining realizable estate shall be dealt with in the following way: the capital, invested in safe securities by my executors, shall constitute a fund, the interest on which shall be annually distributed in the form of prizes to those who, during the preceding year, shall have conferred the greatest benefit on mankind.

—Alfred Nobel, last will and testament[1]

The reading of Alfred Nobel's last will and testament upon his death in 1896 must have been a shock to his relatives. He left 94 percent of his fortune to establishing and annually awarding a set of international prizes, about $250 million at that time. The prizes were to be awarded in the fields of physics, chemistry, physiology or medicine, and literature, as well as to the person who had done the most or the best work for world peace or international disarmament.

1. The full text of Alfred Nobel's will can be found at https://www.nobelprize.org/alfred_nobel/will/will-full.html.

Alfred Nobel was an international industrialist born in 1833 in Stockholm, Sweden, and raised and educated in Saint Petersburg, Russia, where his father, Alfred Nobel Sr., had a weapons factory. As a young man, Alfred worked in that factory, which supplied weaponry to the Russians when they tried to defend their invasion of the Ottoman Turks' territory (in modern-day Romania). In 1853, this conflict became the Crimean War, in which Russia was opposed by Turkey, France, England, and, eventually, Sardinia. The war centered in and on the shores of the Black Sea.

The Nobel factory supplied the Russians with arms, and the czar became interested in Alfred Nobel Sr.'s invention: underwater self-detonating explosives (torpedo mines). The company made investments in production facilities based on product orders from the czar. After the Russians lost the three-year war, the new government did not honor the orders. Nobel Sr.'s enterprise went bankrupt. Back in Sweden, Alfred's mother sustained the family by running a grocery, while he and his brothers rebuilt the company.

An accident in the Swedish Nobel factory in 1864 killed five people, including Alfred's younger brother Emil, and slightly injured Alfred. The incident sparked Alfred's quest for a more stable explosive. Key to his work was nitroglycerine, a highly unstable liquid that had been invented in 1847 by the Italian Ascanio Sobrero. (Sobrero had warned against its use, in fact keeping his discovery secret for a decade.) Through extremely careful research, Alfred found that diatomaceous earth acted as an absorbent to sufficiently stabilize nitroglycerine. He patented dynamite, and Nobel munitions factories popped up in France, Scotland, Italy, and Germany, and more were opened in America and Sweden. He also invented ballistite, a smokeless gunpowder that allowed for rapid fire without having to pause for gun smoke in the air to clear, and he developed new ways to prevent overheating of machine guns, permitting prolonged rapid firing without the need to pause for cooling.

Some of Alfred Nobel's less-famous 355 patents, which nonetheless significantly contributed to the deadly effectiveness of warfare, included improvement in the use of explosives in torpedoes, a pressure regulator for cannons, methods to counteract rifle recoil and dampen sound, improvements in bullet cartridges, and several war rockets and improvements on their velocity. The Nobel organization had established ninety armaments factories by the time of his death.

Nobel's inventions were applied to mining and mountain-blasting activities in addition to war use. Alfred engaged in extensive personal correspondence about war and peace, but it is unlikely that he was a pacifist, as some biographies maintain. Reports that he intended his creations to be used primarily in peaceful activities are not supported by the patent documents, which explicitly state the weaponry nature of his most profitable inventions. According to an article by the Nobel Foundation head of information, Alfred Nobel envisioned that peace would be possible only by development of weapons so effective that opposing armies could mutually annihilate each other. Nobel hoped this thought would horrify the armies, prompting them to disband, and peace would reign.

The idea for Nobel Prizes is alleged to have originated in 1888, upon the death of Alfred's brother Ludvig from heart disease. According to many biographies, a French newspaper mistakenly thought it was Alfred who had died and ran his obituary with the headline "The Merchant of Death Is Dead." This was supposed to have given Alfred the resolve to fashion a more favorable legacy for himself. However, the story cannot be verified as the headline is ascribed to "a French newspaper" with no actual source or author identified.

A poem Alfred Nobel wrote at the age of eighteen refers to "convulsions" he may have had as an infant, and he was described as a sickly child. He suffered from migraine headaches as an adult. In his later years, he was said to have developed chest pains (angina) due to blocked heart arteries. The treatment for angina then was the same as it is today: tiny doses of nitroglycerine, which is known to dilate blood vessels, resulting in a splitting headache as a side effect. Nobel no doubt saw the irony of being prescribed the same basic stuff that he used in dynamite. He refused the medication due to his headaches. He suffered a stroke and died in 1896 at the age of sixty-three. This was the same cause of death that had ended his father's life at the age of seventy-one.

Aside from being a shrewd businessman, an inventor, and a chemist, Alfred also had a great interest in medicine. This may have originated in his family's tendency to develop premature cardiovascular disease—the father and at least two sons were afflicted. He lavished donations on research departments at the Karolinska Institute, a medical university based in Stockholm, Sweden, and he funded Ivan Pavlov's laboratory in Russia. Pavlov's known work at the time was studying dog salivation and collection of gastric juices for production of a popular digestion remedy, but Pavlov was also already well into his experiments on controlling human behavior.

Nobel's will stipulated that there would be annual prizes in the fields of physics, chemistry, literature, peace and disarmament, and "medicine or physiology." The medicine prize was to be awarded for clinically applicable medicine that directly benefits patients as well as basic biological discoveries that broaden our scope of understanding. The awards have been a mix of both.

Many awardees have been recognized for work they did much earlier, sometimes decades earlier. For example, when Ilya Mechnikov received the 1908 Nobel Prize for discovering how certain cells in the body engulf and digest harmful bacteria, it was for work he had done in 1884. When Willem Einthoven got his prize in 1924, it was for the electrocardiogram (EKG) machine he had invented in 1901. The first version of it took up two rooms and weighed nearly six hundred pounds. It required advances in technology and the test of time to witness how indispensable the EKG machine was to become. Currently, EKGs are used countless times per year to detect abnormal heart rhythms and diagnose heart attacks.

Although the Nobel Foundation specifies that the prize is not to be given for lifetime achievements, many have indeed been awarded for a whole body of work, such as the 1905 prize to Robert Koch.

The amount of prize money awarded per year depends on the finances of the private, nonprofit Nobel Foundation as determined by the performance of the investments managed. The award amount in 1901 was 1.5 million Swedish kronor, about 17,738 US dollars at the time. This was roughly twenty times the average annual salary of a university professor. In 2014, the medicine prize was 8 million Swedish kronor, or 1.2 million US dollars.

The deciding body for the Nobel Prize in Medicine is a fifty-member Nobel Assembly chosen from the staff of the Karolinska Institute, a large medical school and research center in Stockholm, Sweden. The assembly, in turn, elects five of its own to serve on the Nobel Committee, each serving two- to three-year terms.

Nomination for a prize is by invitation only. For medicine, the Nobel Assembly at the Karolinska Institute invites nominations each year from members of academies, as well as university professors, scientists, and previous Nobel laureates. In the first year of the prizes in 1901, 128 nominations were received for medicine. Many nominees from that first year would go on to receive prizes in later years, while some names were never seen again. For comparison, in 2015 the assembly sent over 3,000 invitations to nominate; 357 nominations were received and all but 57 had been nominated at least once before.

The committee does not in any way determine the winner by how many nominations it receives. It invites international experts to help it evaluate the work of only the nominees it considers to be top contenders. When it completes its assessments, the five-member committee recommends the winner or winners to the fifty-person assembly. On the first Monday in October the assembly votes on the committee recommendations. A press conference immediately follows the vote, and the awards ceremonies are held in December.

There is nothing inherently "fair" about the process, but then again, the Nobel Foundation makes no claim to fairness. Some very worthy individuals die before the importance of their discovery is appreciated by their colleagues or realized in the broader society. For example, British surgeon Joseph Lister was the pioneer of antiseptic surgery. (Listerine mouthwash was named in his honor.) He had been nominated numerous times by many different people, but his name never made it past the committee by the time he died in 1912.

Some prize decisions were made even in the absence of solid scientific data. For example, in 1927 Julius Wagner-Jauregg was awarded the Nobel Prize for discovering a "cure" for syphilis. Wagner-Jauregg infused the blood of malaria victims into patients who had developed dementia from advanced syphilis infection. His theory was that the fever from malaria would kill the syphilis bug. In his initial 1917 report, Wagner-Jauregg claimed that six out of nine patients "recovered" from syphilitic dementia, for which there was no proof given. Nevertheless, a worldwide sensation was born as others rushed to apply the treatment. It was soon proved useless. The fever therapy competed with an antisyphilitic drug invented by another Nobel winner, Paul Ehrlich, whose drug Salvarsan also probably harmed as much as it helped because it was made of arsenic.

The first fifty years of the Nobel Prize recognized striking developments in our understanding of the human body and its mechanisms of health and disease. The research of Nobel laureates led to indispensable medicines that saved millions of lives, such as sulfa antibiotics and penicillin, as well as discovering hormones that could be lifesaving (insulin, cortisone, thyroid). The development of accurate blood typing, aseptic procedures during surgery, and a method of sewing blood vessels all combined to make surgery a lifesaving option rather than the likely death sentence it had been. These years saw the documentation of many elements of the immune system, detailed work on genetics, discoveries of both electrical and chemical transmission of nerve impulses, identification of vitamins, ascertainment of the insect vectors of malaria and typhus, invention of the EKG machine and cerebral angiography, and great strides in the understanding of biochemistry at the cellular level, respiration, heart rate, eye optics, balance, and muscle physiology. Many times the brainiacs got it right, and the boneheads sometimes not, but often there was a mix of the two qualities.

Enjoy the first fifty years.

Nobel Prize Rules

- No posthumous awards. Initially, prizes could be awarded if the nomination at least was made while the person was alive, but before the winner was decided. It was later changed so that the person had to be named as an awardee while still alive (in October), and would be recognized as the winner even if they died before the December ceremonies.
- No self-nominating.
- Prizes have to be based on published works (except the Peace Prize).
- Committee members can include "foreigners" (meaning non-Swedes).
- No protest can be lodged. In fact, protests and even lawsuits are lodged but they do not influence the committee to change the awards.
- Initially, all minutes and records remained sealed. This was later changed so that records can be opened for purposes of historical research after fifty years have elapsed.
- No more than three people can share a prize per year per subject.
- Not just anyone can nominate a person for the prize. Nominations for the prize in medicine or physiology are solicited by invitation only from:
    - members of the Nobel Assembly at Karolinska Institute;
    - Swedish and foreign members of the medicine and biology classes of the Royal Swedish Academy of Sciences;
    - Nobel laureates in physiology or medicine and chemistry;
    - members of the Nobel Committee not otherwise qualified per the above;

- holders of established posts as full professors at the faculties of medicine in Sweden and holders of similar posts at the faculties of medicine or similar institutions in Denmark, Finland, Iceland, and Norway;
- holders of similar posts at no fewer than six other faculties of medicine at universities around the world, selected by the Nobel Assembly, with a view to ensuring the appropriate distribution of the task among various countries; and scientists whom the Nobel Assembly may otherwise see fit to approach.

Prizes not given

Some years, the Nobel Foundation did not award a Nobel Prize for lack of worthy nominees. Some years there are clearly other concerns that interfere with the foundation's work, such as international war. In the first quarter century of the Nobel Prize, no awards were given in medicine in 1915, 1916, 1917, 1918, 1921, and 1925.

1

The First Nobel Prize

The very first Nobel Prize in Physiology or Medicine was awarded to Emil Adolf von Behring in 1901 for developing a treatment for diphtheria. He was born in Prussia in the former German Empire, in an area that is now part of Poland.

Von Behring received his medical training at the Prussian Army's medical college. He was also the first of many Nobel laureates whose work was driven by the medical problems of warfare. A huge concern to governments at war was infectious diseases, as it was typical that far more soldiers died from illness than from direct battle injury. Even the injured ones who survived initial trauma commonly went on to die of wound infections. One way or another, microbes were responsible for more war fatalities than anything else.

In the late 1800s, von Behring and his Japanese research colleague, Kitasato Shibasaburo, discovered that something circulating in the blood of people and animals with diphtheria somehow helped them fight the disease. Diphtheria is a bacterial infection that gives the sufferer a severe sore throat and causes a thick leatherlike membrane to form in the back of the throat. (*Diphtheria* comes from the Greek word for "leather.") It can cause a person to struggle for breath and eventually suffocate; in von Behring's day, more than half of the soldiers who came down with diphtheria died from it.

Von Behring and Shibasaburo filtered the blood of diphtheria-stricken animals and isolated a substance they named *antitoxin*. The antitoxin helped infected animals recover. Von Behring and another colleague, Paul Ehrlich, injected antitoxin into healthy animals, which successfully allowed them to resist diphtheria infection. So it appeared that antitoxin could both treat and prevent illness. The use of antitoxin reduced soldier death rates from diphtheria from 50 percent to 25 percent.

Another first with this Nobel Prize was the tradition of ignoring the contributions of colleagues who were often codiscoverers and who, in many cases, actually conceived of and performed a great deal of the experimental work. In those days, a Japanese national was given the same disrespect as a "mere woman" or anyone of a less desirable race or skin color. In fact, a year before he came to work with von Behring, Shibasaburo had discovered the bacteria that caused tetanus and found that it was the tetanus toxin produced by the bacteria that gave the symptoms of disease. Von Behring and Shibasaburo jointly published a paper describing their experiments on animals using tetanus toxoid and worked together on every step of the groundbreaking diphtheria research, but when the diphtheria paper was published, it carried only von Behring's name.

The records of the Nobel Committee were made available fifty years later in accordance with the Nobel Foundation rules of secrecy. They show that while twelve nominations were sent in for von Behring, only one nominator also named Shibasaburo. The first Japanese Nobel winner in medicine was not until 1987, when Susumu Tonegawa won for his research in a related subject, antibodies.

It was von Behring's coworker Paul Ehrlich who did the major work of translating these discoveries into a standard medical product that could be commercially manufactured. Both von Behring and Ehrlich were approached by the Hoechst pharmaceutical company, which offered to pay them for putting their discoveries into large-scale production. Von Behring is reported to have wrangled the deal, arranging most of the profits for himself and leaving Ehrlich with only a trivial share.

Two years after the Nobel, von Behring formed the first biotech startup—the Behring Works (Behringwerke)—which made him wealthy. Von Behring outraged the posh citizenry of Capri, where he had a vast estate, because in 1903, he proposed to convert his land on the idyllic island into a vaccine-manufacturing facility as well as a sanatorium for tuberculosis patients. The public outcry caused him to reconsider, and Behringwerke ended up remaining in Marburg, Germany. Today Behringwerke Industrial Park is a biotechnology center employing over five thousand people working in sixteen companies. Behringwerke itself has been absorbed by the international company CSL Limited.

American laboratories soon took up antitoxin manufacture, but not all of them followed the same exacting techniques that brought success in Europe. In fact, often it was crude local public health department facilities that concocted the vaccines. When thirteen St. Louis children who had been inoculated with diphtheria antitoxin died of tetanus in 1902, it was found that the preparation they had been given was made with serum from a horse that had tetanus. This caused the city to fire its top doctor of public health. The deadly event was a big factor in Congress passing the Biologics Control Act of 1902 (also known as the Virus-Toxin Law), which called for regulations to ensure the purity and potency of biological products. That law paved the way for creation of the modern Food and Drug Administration (FDA).

The *D* in the DTaP shot given to toddlers contains a purified version of the diphtheria antitoxin. Diphtheria is a disease we do not see much anymore, but even in von Behring's day, diphtheria was largely limited to conditions of overcrowding, poor sanitation, and the battlefield. His initial interest had been to reduce mortality from infectious disease among the military, where substandard sanitation and overcrowding were routine. In fact, the death rate in the general (nonmilitary) population from diphtheria had dropped dramatically from about 1860 to about 1882. The diphtheria death rate then rose for one decade, dropped steadily, and continued on this decline well before widespread vaccinations were adopted. And improvements in environmental prevention (mainly relieving overcrowding) continued simultaneous to implementation of widespread vaccination, so it is difficult to know with any certainty the degree to which vaccines contributed to this disease not being prevalent today.

Von Behring was reported to have suffered spells of severe depression that caused him to periodically take refuge in so-called sanatoriums. The image of early-twentieth-century madhouses with chained and abused inmates is probably not what von Behring experienced; instead, it is likely he checked into more of a resort-like sanatorium. One such spell garnered headlines in the *New York Times* in 1907. "Behring Denies That He Is Insane" was followed the next month with "Prof. Von Behring Has Recovered." The second news piece read:

> NAPLES, March 19. Signor Spinola, brother-in-law of Prof. Emil von Behring, the eminent German scientist, is authority for the statement that Prof. von Behring has entirely recovered from the temporary mental trouble from which he has been suffering recently as a result of overwork.[1]

However, the official Nobel website reports that von Behring was in a mental institution from 1907 to 1910.

1. "Prof. Von Behring Has Recovered," *New York Times*, March 19, 1907.

In those days, cancer and heart disease were definite medical concerns, but the greater threat to surviving into old age was infectious disease. Von Behring worked extensively to find a preventive or cure for tuberculosis (TB). Even great scientists can be wrong, and in his case von Behring was convinced that human TB was transmitted by bacteria in milk. We now know the typical TB that humans get is transmitted from human to human by respiratory droplets. Von Behring advocated adding formaldehyde to milk to kill the TB bacteria—a truly horrible, poisonous solution that was rejected simply because it made the milk smell peculiar. At the age of fifty, von Behring contracted tuberculosis and eventually died of it when he was sixty-three, although some reports say the cause of death was simply pneumonia.

2

The Parasite and The Pest

The 1902 Nobel Prize was won by Ronald Ross for his discoveries about how malaria was transmitted. *Malaria* is an Italian word that means "bad air" (*mala* "bad," + *aria* "air"), because at one time the illness was thought to be caused by foul air in swamplands.

Humphrey Bogart in *The African Queen* realistically depicted malarial attacks. The victim goes through three stages: the cold stage, the hot stage, and the sweating stage. Shaking chills last for up to an hour, followed by a high fever (106°F or higher) lasting up to six hours, and then profuse sweating for up to four hours. Symptoms include headache, vomiting, delirium, anxiety, and restlessness. All the symptoms go away when the temperature comes down. In some kinds of malaria, the cycle is repeated every forty-eight hours; in other kinds of malaria, the cycle repeats every seventy-two hours. While plenty of people recover spontaneously, untreated malaria can lead to complications and death.

Previous researchers, including Charles Louis Alphonse Laveran, Patrick Manson, and Albert Freeman Africanus King, had proposed that mosquitoes were the vector that transmitted malaria. It is a mystery why Ross received his Nobel in 1902 for this discovery, while Laveran had to wait until 1907 to be recognized for the earlier discovery of the actual parasite.

Ronald Ross was a British citizen born in India, where his father was a general in the British Indian Army. The young Ross liked to write poetry and plays. He worked at musical composition and did mathematics for fun, but his father pushed him toward medical school. Once he became a physician, Ross was inspired by Patrick Manson, who suggested to Ross that he pursue the study of mosquitoes as the malarial vector (the cause of carrying malaria).

In the late 1800s, Ross was assigned as a medical officer in India, a post that he sought because it allowed him to be in malaria-prone regions. He let a mosquito feed on the arm of a malaria victim and then dissected the insect. He discovered a cystic form of a parasite within the stomach of the mosquito and concluded that this could be the causative organism of malaria. Before Ross could carry out definitive experiments to prove this theory, he was transferred to a town where malaria was not common. His mentor, Dr. Manson, pulled some strings back in England and was able to get Ross reassigned to a swampy area where malaria was rife. In fact, Ross himself came down with malaria but recovered to continue his studies for another two years.

Ross found it difficult to get good experimental subjects because they were medicated almost as soon as they fell ill, so he resorted to experimenting on birds. He was able to trace the life cycle of a parasite from an infected bird to a mosquito and identified the mosquito as gray or brown and "dappled-winged." This research won him the prize.

However, Ross did not identify the exact species of mosquito in his initial work, not being an insect expert. His experiments were limited to birds, not humans, and he did not prove that the parasite he found could cause malaria in humans. In fact, what he discovered was a different parasite specific to birds that does not infect humans at all.

Meanwhile, an Italian research team led by zoologist and physician Giovanni Battista Grassi was also studying malaria, a grave problem in coastal Italy at that time. Grassi was a serious student of insects of all kinds. He studied worms, parasites, and termites, and even discovered a new species of spider that he lovingly named after his wife.

Grassi and his team correctly identified the parasite-carrying mosquito as being the *Anopheles* species, described the complete life cycles of the three malaria parasites that infect humans, and demonstrated that the parasites go through mosquitoes and into humans. Furthermore, they established that it was only the female of the species that transmitted the parasite.

The Nobel Committee initially intended to award the prize jointly to Grassi and Ross, but before that could happen, Ross attacked Grassi as a fraud publicly and in professional publications. In severely inflammatory language of the time, he called Grassi a mountebank (a person who sells fake medicine by telling elaborate lies), a cheap crook, and, uncreatively, a parasite. Ross's protests seemed excessive, and it is no wonder: the ferocity of his attack related to a failed attempt at spying on Grassi's work. Ross had sent a colleague, Dr. Thomas Edmonston Charles, to make an informal visit to Italy, casually poke around Grassi's lab, and make friendly inquiries; he was to report back to Ross with the scoop. Grassi correctly suspected the reason for Charles's visit and apologetically explained that he and his team had not progressed much in their research. Two weeks later, the Grassi team published its findings, enraging Ross.

The Nobel Committee turned to the renowned German scientist Robert Koch to be an arbitrator. By this time in 1902, Koch was already famous for being the "father of clinical microbiology." In the late 1890s he had gone to Italy to study malaria and interacted with Grassi. Grassi had let Koch know that he disapproved of the German scientist's analytical methods, so Koch was not a neutral party at all. He sided strongly with Ross, and despite twenty-one nominations, Grassi was left out of the prize entirely.

According to contemporary reports, Ross was a chronically unhappy, arrogant man. In his Nobel acceptance speech, he omitted giving any credit to his former mentor, Dr. Manson. When Ross returned to London, he was offended that Manson recommended him for a position at the Liverpool School of Tropical Medicine rather than the prestigious London School. Ross chronically complained about being underpaid by the Liverpool School and quit twice in protest; he was ultimately fired without any pension.

Ross also failed to build a sustainable private medical practice, while Manson enjoyed popularity and financial success as a private physician. In 1930—eight years after Manson's death—Ross wrote the book *Memories of Sir Patrick Manson*, in which he blatantly minimized Manson's work on malaria and denied Manson's influence on his own malaria work.

Ross was variously described by his contemporaries as generating professional hostility and being chronically maladjusted, impulsive, and a "tortured man." He was said to be quick to take offense and capable of magnifying a petty affair out of all proportions. By 1928, he had brought four libel suits against supposed offenders.

The discoveries of the malaria parasite did not immediately lead to an effective remedy. Quinine, from the bark of the cinchona tree, is a malaria remedy dating back to at least the seventeenth century in Europe. Before that, it was used medicinally for centuries in South America. (Quinine is also the flavoring of the "tonic" in gin and tonic.) Quinine and its chemical derivatives are still effective in some parts of the world today, but parasite resistance has emerged in many places. Antibiotics and newer drugs are being used, but they cannot keep pace with how rapidly the parasite develops new drug resistance.

Prevention consists primarily of draining swampy areas and instituting mosquito control, secondarily of using insecticides, and finally of prescribing preventive medications. Since 2010, the Bill & Melinda Gates Foundation has been distributing insecticide-impregnated mosquito nets to the ninety-nine countries in which malaria is prevalent.

In the 1960s, the World Health Organization (WHO) led a campaign to eliminate malaria, but its early efforts were not sustained once the initial funding sources dried up. This created an even bigger problem. The WHO campaign drastically reduced malaria for about a generation but did not continue the effective control measures, so malaria returned with a vengeance. The younger generations had never been exposed and were especially vulnerable due to lack of immunity, and resultant death rates were even higher than before.

For example, on the main island of Sri Lanka, the disease had been almost eradicated several decades ago. With no funding for continuation of the programs, malaria now infects ten thousand Sri Lanka islanders per year. WHO estimates that worldwide there are 300 to 500 million cases of malaria with about a million deaths per year.

3

A Bright Future

The 1903 Nobel Prize went to Niels Ryberg Finsen for the application of light therapy to treat skin lesions of tuberculosis (TB). Finsen had a rough start in life. He was born in 1860 on the Faroe Islands—a treeless, relentlessly wind-raked land directly north of Scotland, in between Norway and Iceland. His mother died when he was only six. On behalf of the Danish government, his father collected taxes from sheepherders; sheepherding was the main industry on the islands. Early on, Finsen was a profound disappointment in school. The principal at his Danish boarding high school described him as "a boy of good heart but low skills and energy." He was transferred to a school in Iceland, where his grades improved, and then he defied expectations by getting through medical school.

Sunlight therapy was one of the first-known medical treatments, and evidence of its use comes from the ancient Babylonians, Assyrians, and Greeks. The Indian ayurvedic healing system assigns specific colors to each of the seven chakras, representing internal organs as well as seats of emotional responses and roughly correlating to what is known today as the endocrine (hormone) system. The Persian philosopher and physician Avicenna of the tenth to eleventh century AD advocated herbs of certain colors for various conditions. The sixteenth-century physician Paracelsus recognized the importance of light and color in maintaining health and treating disease.

There was renewed scientific interest in light and color therapy in the nineteenth century, and two prominent books on the subject were published in the 1870s when Finsen was a teenager. A. J. Pleasanton, a retired Civil War militia general, conducted experiments on the effects of blue light on the growth of plants and the healing of wounds in animals, all described in his book *The Influence of the Blue Ray of the Sunlight*. In 1878, English mathematician and philosopher Edwin Babbitt published *The Principles of*

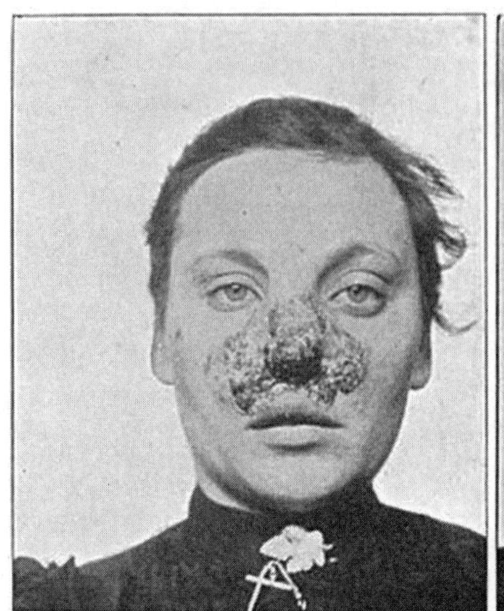
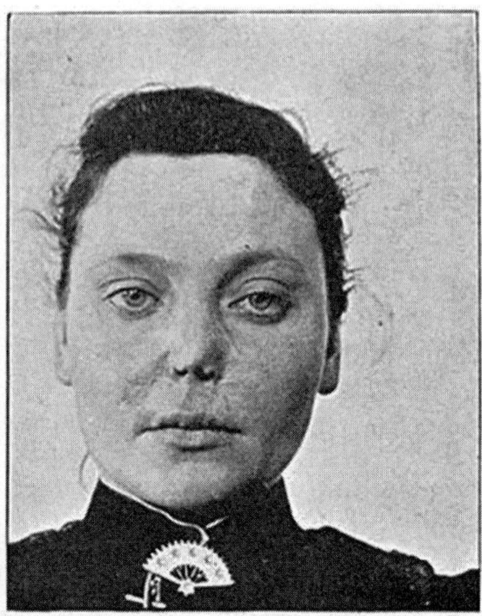

Finsen's light therapy: Before and after photographs of a patient suffering from *lupus vulgaris* (tuberculosis of the skin)

Light and Color, in which he described light as both an energy wave and a particle, predicting the modern physics theories of Einstein some fifty years later. Babbitt made a device that divided a light source into the color spectrum and then focused the rays of one color on a body part for specific treatment.

Finsen had personally experienced the beneficial effects of light by treating his own symptoms of weakness and anemia with sunbathing. He sought to improve on the known general benefits of sunlight to develop specific therapy for diseases. The sores caused by the bacterial infections smallpox and tuberculosis could be profoundly disfiguring, especially when they occurred on the face, and there was no effective treatment at the time. Finsen worked to identify the colors, or light fractions, that were most effective for each condition.

First, Finsen found that the ultraviolet (UV) fraction of light made smallpox sores worse. (The UV wave is perceived as bluish by the eye.) He filtered that out and was left with a light that appears red to the eye; this red light sped up the healing of smallpox lesions. But smallpox was already rapidly on the wane and was much less of a public health concern than tuberculosis. When Finsen tried the ultraviolet blue light on the skin sores of TB, they healed rapidly. He developed a special lamp to deliver the correct wavelength, and it became a foundation of TB therapy. Residential resort–like "Finsen Institutes" for sunbathing and lamp treatment popped up all over Europe and Russia, usually in mountainous regions to be closer to the sun.

Finsen is hailed as the founder of phototherapy in dermatology. In 1886, he founded the Finsen Light Institute in Denmark, later taken over by the government university medical center. In 1904, a review of over twelve hundred cases showed 51 percent cured and 35 percent markedly improved. A letter published in the January 1902 *Journal of the American Medical Association* describes the results of Finsen's light therapy: "The transfiguration truly borders upon the marvelous. The once repulsive features are changed into those of a normal being."[1]

Light institutes were established around the world and became instantly inundated with patients. Interest in phototherapy went beyond the treatment of the skin. Refinements of light therapy were developed to treat festering wounds. For other conditions, including bloodstream infections and cancer, some researchers had success in extracting a small amount of blood, treating it with UV light, and returning it to the patient. But soon there was declining interest in Finsen's discoveries, probably because of the natural decline in TB cases. In the absence of any specific drug treatment, the incidence of TB nevertheless dropped rapidly throughout the nineteenth century. At the same time, it was gradually becoming a less deadly disease: by 1890, the death rate from TB was half of what it had been in 1821. By the time the antibiotic era started—with the discovery of spectromycin in 1940—TB was no longer a major public health problem.

After Finsen, the next notable color-therapy activity was from an ayurvedic physician who was a naturalized American citizen from India, Dinshah Ghadiali. His ayurvedic training was not recognized in the States, but he went on to become a chiropractor and naturopath and founded the National Association of Drugless Practitioners in 1912. Today this might seem to be a nonthreatening act, but in those times, the American Medical Association (AMA) was still trying to achieve dominance in the health-care field. The AMA had to fend off homeopaths, naturopaths, herbalists, and chiropractors by asserting that there was only one legitimate model of disease management and it was based on drugs, surgery, and, later, irradiation to the exclusion of anything else.

In 1920, Ghadiali announced the results of years of work to translate the color principles of ayurveda into a Western application. He called it

1. A. K. Warner, "A Visit to Finsen's Institute," *Journal of the American Medical Association* 38, no. 3 (1902): 188.

spectro-chrome therapy—a method of projecting color beams of light on the body of the patient. Within a mere twenty-five years of Finsen's Nobel Prize for color therapy, the considerable forces of the AMA were hard at work to criminalize the treatment. Ghadiali was accused of fraud in 1931 for daring to claim that color therapy was medically effective. The court transcript included testimony by three well-respected physicians who were using color therapy, including Dr. Kate Baldwin. Dr. Baldwin had graduated from medical school in 1890, around the time of Finsen's light experiments. After practicing as a physician and surgeon for decades, Baldwin wrote about her experiences with color therapy in a paper presented to the Pennsylvania State Medical Society:

> For about six years I have given close attention to the action of colors in restoring the body functions, and I am perfectly honest in saying that, after nearly thirty-seven years of active hospital and private practice in medicine and surgery, I can produce quicker and more accurate results with colors than with any or all other methods combined—and with less strain on the patient. In many cases, the functions have been restored after the classical remedies have failed. Of course, surgery is necessary in some cases, but the results will be quicker and better if color is used before and after operation.[2]

Physicians testified that they used light therapy to treat such diverse conditions as kidney failure, cancer, pneumonia, gonorrhea, wound infections, eye problems, and third-degree burns. The testimony of Baldwin and other physicians temporarily held off the regulators. Ghadiali was acquitted and narrowly avoided being deported. In 1945, he was again accused of false claims, resulting in the seizure of his books and colored light bulbs. He avoided prison only by agreeing under duress to abandon spectro-chrome therapy. The Food and Drug Administration (FDA) officially outlawed his light device in 1958. Currently, the FDA sanctions color therapy only in very limited applications for four conditions. Ultraviolet light therapy is a standard psoriasis treatment, and UV is used for yellow jaundice in newborns and for vitiligo (white patches caused by loss of pigment). Sunlight or full-spectrum light boxes are accepted treatments for seasonal affective disorder (SAD). Using a therapy for something other than what the FDA approved it for is called "off-label prescribing." It is not illegal but can leave the practitioner open to a malpractice lawsuit based on insufficient informed consent. The FDA requires alternative practitio-

2. Kate Baldwin, "The Therapeutic Value of Light and Color," *Atlantic Medical Journal* (April 1927): 431–32.

ners offering light therapy to disclose that they are "not intending to treat disease."

There has been a renewed interest in the scientific documentation of the effects of light in biological systems in the last twenty years. Today it is well known that light modulates biochemical reactions in bacteria and in mammals, including humans. Sunlight has been shown to reduce inflammation of liver cells, suggesting that it could be useful in some cases of hepatitis. Sunlight directly affects a type of immune cell called *macrophages*, which in turn activate vitamin D receptors and stabilize the immune reaction of the body. Ultraviolet A (UVA) light effectively treats autoimmune skin disorders such as atopic dermatitis and eczema. It alters the oxygen within cells, triggering the death of excessively active immune cells. UVA light causes rapid dying off of human leukemia cells. Visible blue light kills off methicillin-resistant *Staphylococcus aureus* (the notorious antibiotic-resistant flesh-eating bacterium). Blue light attacks the bacteria responsible for acne boils and is being studied for its effects on the bacteria that cause stomach ulcers. Blue light has been shown to reduce organ injury from an episode of low blood flow such as occurs with heart attack and stroke. Visible blue light affects abnormal tissue while leaving normal areas alone, meaning zero side effects for the patient. This is different from bluish ultraviolet light that could have some harmful side effects on normal tissues.

Recent studies demonstrate that purple (violet) light and green light kill off bacteria that cause dental gum disease. Violet light has been shown to interrupt a herpes infection after the lesions are painted with a red dye, but it seems to work only if applied in the first two days of the rash. There is an improved survival rate from carbon monoxide poisoning after treating the lungs with colors ranging from yellowish green to orange to reddish orange. Visible red and near-infrared light are being studied to attack fungal infections, promote bone healing and wound healing, and limit or reverse scar formation. A more advanced use of phototherapy is to inject tiny fat bubbles carrying a cargo of chemotherapy drugs; once they reach their destination they are activated with infrared light, causing them to dump their contents directly into the tumor.

The vested interests of the AMA, backed by the regulatory power of the FDA, caused the loss of a century of research into an entire field of effective therapies sparked by Finsen's discoveries. We have only begun to emerge from the dark in the last couple of decades, and light and color therapy has once again been subjected to rigorous scientific investigation. We finally have a robust body of data on the molecular and biological basis of over a thousand years of empirical observations: color therapy works. Despite this, a typical patient is offered color therapy only by an alternative practitioner or by figuring out self-treatment from an internet search.

Young Niels Ryberg Finsen never enjoyed robust health, but it is unclear what he suffered from. Some reports describe that he began to get the symptoms of Niemann-Pick disease when he was in his twenties. That is a rare, genetically acquired defect of fat metabolism that causes brain degeneration and dementia. It is doubtful Finsen could have become a Nobel Prize–winning medical researcher if he had had this disease. Other reports describe Finsen having a constrictive heart disease (the lining of the heart stiffens and thereby limits heart pumping). Finsen himself thought his heart failure was a residual of tapeworm infection contracted in his youth in Iceland, and he was probably right. He received news of his Nobel Prize nomination while confined to a wheelchair and died the next year at the age of only forty-three, leaving behind the Finsen Institute to carry on his work.

The Finsen Institute gradually moved into the field of ionizing radiation research and treatment. As of 2017, the Finsen Center at the University of Copenhagen specializes in the treatment of cancer, infectious diseases, and blood disorders but does not heavily support research into light therapy.

4

Science Promotes a Slave State

Ivan Petrovich Pavlov won the 1904 Nobel Prize for his discovery of the nerves that control the stomach glands and their production of digestive fluids. Pavlov was born in Russia to a family with a rich heritage of religion. His father served as the village priest, and his grandfather had been a sexton. Pavlov started off in seminary school but abandoned theology for medical studies. After medical school, he studied with a heart specialist in Leipzig, Germany.

The chair of the department of philosophy at the University of Leipzig at that time was Wilhelm Wundt, the first person in the world to call himself a psychologist. Strictly speaking, the word means "one who studies the soul," but Wundt asserted that humans are devoid of a spirit and have no self-determinism. It is unknown how Wundt and Pavlov interacted during his stay at Leipzig, but there is no doubt that they crossed paths and followed each other's work. Wundt promoted the ideas that learning was based solely on nerve pathways and that human behavior would eventually be entirely explained by the anatomy of the body. This belief became the foundation of Pavlov's later experiments.

Pavlov completed his postgraduate studies and returned to Saint Petersburg, Russia, where he was appointed head of the Physiology Department at the newly established Imperial Institute of Experimental Medicine. In 1883, after Pavlov had led the new institute for three years, it received a generous donation from Alfred Nobel. (The Nobel family had lived in St. Petersburg for several years, and it is where Nobel received his education as a youth.)

Pavlov experimented on live animals to determine the function of nerves with an initial focus on the nerve control of the heart and circulation. He thought he had discovered that four main nerves regulate heart function, an idea that was soon proved false. Pavlov had better powers of observation when he studied the digestive system of dogs; there he discovered that the nerves of the pancreas control the flow of secretions of digestive substances. This was partially true, but it is now known that nerve control is the not the *only* factor affecting how and when the pancreas secretes insulin.

In 1889, Pavlov discovered which nerves control the stomach glands and their production of digestive fluids. For this work, Pavlov would receive the 1904 Nobel Prize. To measure the stomach juices, he devised a surgical operation in which he diverted the flow of digestive fluid of dogs into a pouch outside the body. Selling the dog digestive fluid to treat indigestion became a lucrative side business, significantly augmenting his income and increasing his laboratory budget by 70 percent. It was such a popular remedy for stomach ailments in Russia, France, and Germany that Pavlov set up a factory in which each large factory dog would produce about a liter of the precious digestive liquid in a day. The dogs had been operated on to install a hose into their swallowing tubes and then into the stomach, and the hose was connected to a collecting flask outside the body. The dogs were restrained in standing positions at tables, their harnesses held taut through a crossbeam above them. Bowls of fresh meat were put just outside the reach of the hungry animals to force the activation of the glands to pour forth digestive juices.

From about 1901 until the end of his life in 1936, Pavlov turned his efforts entirely toward trying to prove the animal nature of humans. The basic observation that had captivated his attention was noticing that a dog salivates in anticipation of a meal. This is something that any pet owner knows, but Pavlov celebrated it as a great discovery, and it became the cornerstone of his theory of *conditioning*.

According to the voluminous writings of Pavlov, human behavior is entirely hardwired. Pavlov considered human actions as completely determined by external stimuli causing responses in the brain and nerves. In animals, Pavlov thought there was only one system of nerves, but in humans he said there was a second system that responded not only to physical things but also to words. This supposed system of nerves has never been found in a human body.

Theory aside, Pavlov proceeded on a decades-long course of subjecting animals to tortuous experiments to show that their behavior could be controlled by external stimuli. He was able to demonstrate that previously normal dogs could be turned anxious, fearful, or apathetic at the will of the technician. Pavlov and his assistants performed thousands of experiments on dogs and also on cats, monkeys, and, eventually, people.

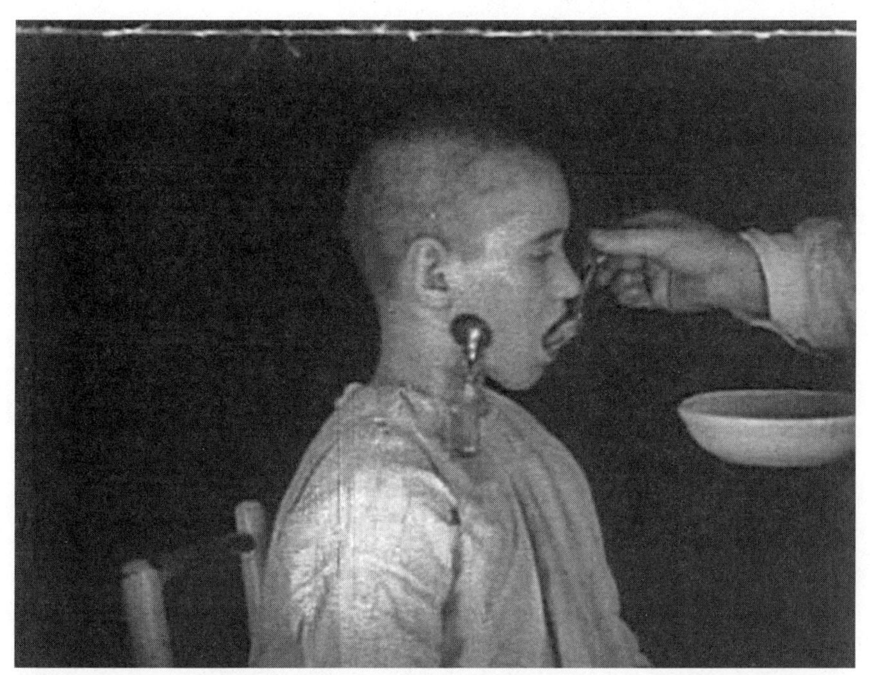

From the Soviet film describing the glories of Pavlovian behavioral conditioning: this child, probably an orphan, had a drain surgically placed in his salivary gland through the cheek, to collect saliva when he was teased with food.

The classic version of Pavlov's famous experiments on conditioning involved setting a metronome to click at every feeding time. Once the dogs were used to this, the metronome clicked but no food was brought. Pavlov noted the dogs salivated nonetheless. This is what brought Pavlov great fame in the world of psychology. Later experiments flipped from positive conditioning, where the stimulus was associated with food, to negative conditioning, where the stimulus was routinely followed by punishment. The metronome signaled that the dog was going to be electric shocked, and it was. Eventually, the dogs shook with terror every time the metronome started, even when there was no follow-up shock.

Pavlov did similar experiments on orphan children, and a few of the photographs and films survive. One photo shows a boy with a hole drilled into the side of his face that is connected to a tube for collecting saliva similar to the operations on Pavlov's factory dogs. An adult is holding a spoon of food near but not at the child's mouth. Another film shows a child of three or so lying on a cot, strapped into headgear, with a tube-like chute suspended directly above the child's face. A small inflatable bulb is strapped onto the child's hand. When the operator causes the bulb to inflate, the chute above the child's face sends cookies rapidly sliding down, aimed at his mouth. The child opens his mouth and gobbles the cookies. After he got used to this, the bulb would inflate even though no cookies came down the shoot. Yet the boy still opened his mouth in anticipation of a flood of cookies.[1] It is unclear what that was supposed to prove.

An example of what Pavlov called *experimental* neurosis is when he showed dogs a picture of an ellipse and one of a circle. When the dogs correctly chose the ellipse, they were fed. Dogs were very good at this differentiation, but then the ellipses were gradually altered to be less elliptical and made to look closer and closer to a true circle. Finally, when the dogs could not differentiate between the two, they displayed extreme anxiety about which to choose. Being wrong meant no food that day.

In a more brutal series of experiments, Pavlov surgically closed off the throat and removed the esophagus from dogs (the body part that connects mouth to stomach). The dogs were kept alive by tube feedings directly into the stomach. He let these dogs get hungry by stopping the tube feedings and then allowed them to try to eat a bowl of food, but the food had nowhere to go, the throat being closed off and connected to nothing. The great scientific outcome of this study was that the hungry dogs kept trying to eat even though the food just spilled back out of their mouths.

Pavlov wrote extensively on the supposed neurological causes of insanity. He said that schizophrenia, for example, was "chronic hypnosis" due to "weak" brain cells. It is known that Pavlov had access to incarcerated mental patients, many if not most of whom were only political dissidents. But the full extent of Pavlov's activities with humans is not known.

Pavlov admitted he had a lifelong problem of trying to prove his theories but certainly succeeded in steadily supplying the Soviet state with various means to subdue the population. The Communists picked up his ideas of using forceful verbal cues and punishment tactics to obtain obedience to an external authority. A subtler Soviet application of Pavlovian concepts was to inundate the population with daily propaganda, pushing the idea that you must think what everyone else thinks. The penalty for not following the party line was lack of food, employment, housing, or being picked up as

1. This film can be seen in *The Brain: A Secret History*, episode 1, "Mind Control," directed by Alicky Sussman, presented by Michael Mosley, http://wn.com/pavlov's_children, which is also available on YouTube as *Ivan Pavlov: Experiments on Conditioning*, https://www.youtube.com/watch?v=N5rXSjId0q4.

a dissident. The eventual goal, according to a psychiatrist writing on mind control, was "Big Brother's voice resounds in all the little brothers."

Pavlov was at the Imperial Institute of Experimental Medicine until his death in 1936, and the institute's archives as well as Pavlov's own writings give a picture of the person. Pavlov was nasty toward dogs and people alike—no real surprise, since he regarded humans as no more than animals. He was prone to violent moods that caused his coworkers to avoid him for days at a time. Pavlov called his outbursts "spontaneous morbid paroxysms." He was an avowed atheist, writing that religion was a coping mechanism for people with weak nervous systems who could not withstand the stimuli of the environment. Pavlov was openly anti-Semitic but balked when the Communist state tried to rid his laboratory of assistants whose fathers were Orthodox Christian priests. (His father had been a priest, and his grandfather was a parish sexton.)

Pavlov was highly critical of Lenin's regime, but he could afford to be. Although the cash winnings from his Nobel Prize were confiscated by the state (as was its policy), his laboratories were well supported and experienced continual expansion. He knew this government did not want such an important mind-control researcher to fall into the hands of a foreign power. It would also have been embarrassing to the image of Soviet prowess if he ever left for a Western country. Pavlov used these facts to demand better funding for his laboratories and privileges and protections for himself and his family.

Pavlov's activities were uninterrupted even during the Red Terror. From 1917 to 1922, Lenin's police systematically selected so-called counterrevolutionaries for torture and execution. This included mass killings of the populations of whole villages, factories, schools, and farms. The targets were "representatives of the overthrown class" (formerly the upper class) as well as any people known or merely suspected of not supporting the People's Revolution.

Pavlov was already an older man in the 1920s when Stalin came to power and by then he had three massive laboratories that employed hundreds of technicians and researchers. He remained unscathed throughout another internal terror campaign: Stalin's Great Purge, which took place from 1934 to 1939. Execution, Siberian exile, or imprisonment for life was dealt out to persons identified as enemies of the Soviet people. Victims were tossed out of the Party, the army, and the intelligentsia, but despite Pavlov's criticisms of the regime, he was not silenced.

Stalin was advised by leading Communist Party officials to tolerate Pavlov's occasional outbursts precisely because his actions spoke louder than his words. "Despite all his grumbling, ideologically (in his works, not in his speeches) he is working for us," wrote Nikolai Bukharin, former editor of the newspaper *Pravda*.

In fact, Pavlov's demonstrations on the various ways to create insanity and use negative conditioning to bring about a whole population of meek, neurotic citizens were strongly supportive of the Soviet goal: the creation of a terrorized citizenry that would not have the courage to revolt. *Pravda* editor Bukharin himself went down in the Great Purge and was executed in 1936, while Pavlov remained untouched.

Pavlov's methods were very effectively translated to population control. Simply put, a lie told often enough becomes a sort of truth, especially with enough reinforcement (food if obeyed, punishment if disobeyed). This was hardly a brilliant Nobel Prize winner earnestly working for the good of humanity, notwithstanding his discovery of a profitable digestive remedy.

5

His Eminence

Robert Heinrich Hermann Koch won the 1905 prize for his investigations and discoveries regarding tuberculosis (TB). Koch was born in Germany in the mid-nineteenth century and educated at the country's best schools. In the tradition of medical researchers of the Western world, Koch's earliest experiments were on himself. His doctoral thesis showed how a high-fat diet caused a certain acid to be found in the urine. In one aspect of the study, he ate a half pound of butter a day for several days and dutifully collected his urine. He took a stab at private medical practice in small-town Germany, and on the side, he went bowling, played the zither, and did some beekeeping. He kept several wild and domestic animals on his property.

An outbreak of anthrax in farm animals prompted his study of the life cycle of the anthrax microorganism. Koch painstakingly documented that anthrax bacteria existed in the soil, in a hardy spore form. There it remained dormant until conditions were good for reproduction, and then it emerged and became capable of infection. His entire study was focused on farm animals, as human infection with anthrax was not a significant public health problem. (Anthrax is a disease of hoofed animals, and rarely a wool handler or sheepherder can get it from direct contact with the open sores or undercooked meat of an infected animal. The dry soil spores of anthrax can be inhaled, but this is a very unusual way to contract the disease.)

Koch took anthrax bacteria from an animal that had died from the disease and made a pure culture of anthrax, then injected the bacteria from the culture into a healthy animal to make it sick with the classic signs of anthrax. He perfected the technique of making a *pure culture* (one that would grow only a single kind of bacteria) on a gel plate of nutrients rather than the usual method of growing a mixed culture of microbes in a flask.

He was the first to take a photograph of a microscopic organism and the first to show the complete life cycle of an infectious bacterium.

Although it seems obvious to us, in 1873 it was controversial to claim that a specific bacterium caused a single disease. The prevailing idea at the time, endorsed by such luminaries as Louis Pasteur, was that bacteria constantly changed form to cause multiple diseases. Instead, Koch advocated the idea of "one bacterium = one disease." Koch was not the first to believe this, but his anthrax study provided the first significant support of the theory.

Europe had recently experienced the war of 1870 (called the Franco-Prussian War by most of the world but called the German-French War in Germany). The contest between countries extended to medicine as well. Koch's theory disagreed with the ideas of the great French biologist Louis Pasteur; Pasteur held his ground but followed Koch's work with interest.

Tuberculosis was an area of highly competitive research because in the nineteenth century, TB was still causing one in seven deaths. With the improvement in microscopes and pure-culture techniques, it was just a matter of time before someone would identify the causative bacterium. In 1882, Koch beat out other efforts and triumphantly announced his discovery of the tubercle bacterium. Pasteur and others doubted his claims because they had not been able to isolate the one Koch described; Koch retorted that they just had sloppy culture techniques and were not using good microscopes. This accusation caused an uproar at the time, but it was probably true.

The German government was happy to emphasize Koch's discovery. He was lauded as a national hero. The government expanded his research institute and gave him research grants, an increase in salary, and more assistants. The Prussian Imperial Health Office held a German Exposition of Hygiene and Public Health and invited the world's scientists to tour the laboratories of the great Dr. Koch.

Meanwhile, another scourge attracted Koch's attention. Cholera was a huge problem, especially in nineteenth-century cities, in which sewage could easily contaminate drinking water. Cholera bacterium produces a toxin that causes severe, watery diarrhea potentially so extreme that it leads within hours to severe dehydration and electrolyte imbalance. Koch's expertise with a microscope led to finding the bacterium that causes cholera, and he was championed the world over as the discoverer.

In those days, however, the Germans, English, and French dominated medical science. In fact, Italian researcher Filippo Pacini had published a

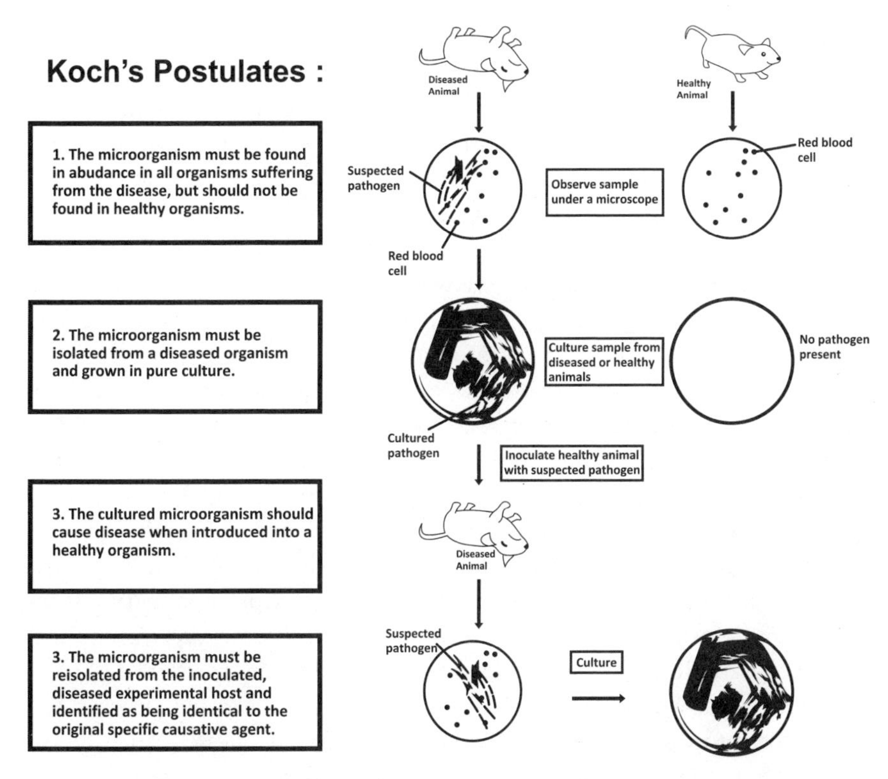

Diagram explaining "Koch's postulates" of infectious disease

paper describing the cholera organism seen through his own microscope some thirty years earlier in 1854 when Koch was only eleven years old. This was either unknown or ignored at the time of Koch's rediscovery, so Koch got all the credit.

Many others had noticed that cholera was associated with contaminated water supplies, but Koch was the first to show the practicality of how to contain it. He conclusively demonstrated that running the municipal water supply through a sand filter could remove the cholera bacteria. This led to major improvements in the sanitation of public water systems the world over.

If he had stopped there, Koch probably would have received the first Nobel Prize instead of the fifth. His subsequent search for a cure for TB did not go so well. In 1890, he concocted a remedy he called tuberculin and injected it under the skin of his own arm. He dutifully noted that he experienced pain and redness at the injection site, joint pain, cough, weakness, fever, shaking chills, nausea, and vomiting. He then gave a much-diluted preparation of tuberculin to the poor patients in a local charity hospital, an accepted practice in those days for unproven treatments. Reliably, those with TB or a past exposure to TB had a pronounced skin reaction while those who did not have TB had almost no reaction. In a great leap of faith, the skin reaction was interpreted as indicating that a cure was taking place.

Koch was a careful researcher, and there is evidence that he knew he did not quite have the necessary studies to announce tuberculin to the world as a valid cure just yet. However, the Prussian government wanted to continue pushing the strong wave of apparent German superiority in medicine. It is probable that Koch was coerced into publishing his findings prematurely.

The world flocked to his institute in Berlin. Newspaper reports of the time told of local citizens being alarmed at the influx of some four thousand germ-spewing TB patients from around the world seeking the cure. A thousand doctors came too, among them Arthur Conan Doyle (he had not yet been knighted), who was first a physician before becoming better known for authoring the Sherlock Holmes novels. Dr. Doyle had no illusions about the shortcomings of modern medicine. Upon graduating from medical school at University of Edinburgh in 1881, he drew a comical sketch of himself receiving his diploma, captioned "Licensed to Kill."

Among the throngs of physicians and laypeople seeking to learn more about the supposed cure for TB, Dr. Doyle wrote of the difficulty "in catching even the most fleeting glimpse of its illustrious discoverer":

> Like the Veiled Prophet, he still remains unseen to any eyes save those of his coworkers and research assistants. The stranger must content himself by looking up at the long grey walls of the Hygiene Museum and knowing that somewhere within them the great master mind is working.[1]

There were two main problems with tuberculin.

In his first paper, Koch did not tell anyone what was in it. This actually directly violated a German law that made it a crime to sell secret remedies. It was a testament to how much Koch was regarded as pretty much *the* authority that he was called a hero for the cure before anyone even knew what it was.

The next little problem was that there had been no long-term studies to actually prove that tuberculin cured anything. Still, the use of tuberculin spread around the world quickly, and soon enough, the numbers started pouring in: it was a total flop.

1. Arthur Conan Doyle, "Dr. Koch and His Cure," *The Review of Reviews* (December 1890): 552.

In 1891, by the time the failure of tuberculin was obvious, Koch revealed his secret formula: tuberculin was no more than dead tubercle bacteria that were ground up and suspended in glycerin. By 1906, doctors discovered that the vigorous skin reaction seen in people already infected with TB was *hypersensitivity*: their bodies were already sensitive to the TB bug, and the skin prick with dead TB bacteria just caused a hypersensitive reaction at the site of the injection. So while tuberculin was not useful for treating TB, it did become the basis for a reliable test to tell if someone had already been infected with TB bacteria. A version of the tuberculin skin test was used in this way for over a hundred years, only recently replaced with a more sensitive blood test.

After the tuberculin embarrassment of 1891, Koch abandoned TB research and dramatically changed his life. He took up with a seventeen-year-old girl and soon ended his twenty-six-year marriage to wed his paramour, even though he was almost thirty years her senior. He traveled the world to work on malaria, cholera, and tick-borne diseases such as sleeping sickness and rinderpest. This took him to India, Egypt, South Africa, New Guinea, Japan, and America. Twice his new wife came down with malaria on these journeys.

In 1901, Koch again became involved in a controversy. In his 1882 paper, he had stated that cow TB and human TB bacteria were one and the same but by 1901 felt certain the bacteria were different. This conflicted with the ideas of Pasteur, who had pioneered the idea of boiling cow's milk to kill bacteria (*pasteurization*). Koch said there was no need for pasteurization since the TB bug in milk did not cause human tuberculosis. A German high commission studied the question for a couple of years and ultimately sided with Koch while other countries continued to pasteurize milk at high temperatures to kill cow TB bacterium.

The pasteurization battle made it awkward to consider Koch for the first-ever Nobel Prize that year. In fact, we now know he was mostly right. The two are indeed subtly different strains of bacteria. Humans can get infected with cow TB, but it is a rare occurrence. And when humans do get cow TB, the disease does not get passed to other humans. In contrast, respiratory droplets pass human TB from person to person. Milk was certainly not the cause of most TB cases in Koch's day.

The irony of this entire hubbub is that tuberculosis was already becoming less prevalent and turning into a less deadly disease throughout the 1800s. In the first ten years of Koch's boyhood there was a steep decline in the TB death rate, and it was still falling steadily through the period of his discoveries and failed treatment. In the 1900s, even the Industrial Revolution and two world wars did not stop the ongoing steady decline of TB. It took another steep decline with the invention of the first anti-TB drug in the 1940s. The only rise in TB in recent times came with the advent of AIDS from the human immunodeficiency virus (HIV), but that increase was nothing of the magnitude seen in the early 1800s. TB incidence started to fall again by 2010.

No single cause of the decline of tuberculosis infections in the pre-drug years has ever been found. However, since most organisms depend on live hosts to propagate, it is not uncommon for an organism that is at first highly deadly to modulate its behavior to become less deadly. It is the idea of the greatest good, and it would not do the tubercle bacterium any good to kill off *all* of its human hosts. Other factors contributing to fewer TB deaths may have been the improved antiseptic conditions in hospitals. As soon as doctors got the idea to wash their hands and wear masks to prevent transmitting disease from patient to patient themselves, it became much safer to interact with the medical profession. Similar steep declines were occurring in scarlet fever, diphtheria, whooping cough, and measles well before any vaccinations were developed.[2]

Koch's institute in Berlin was well funded and attracted the brightest researchers from around the world. In fact, most major discoveries in infectious disease were made there or under German influence. At the height of his popularity, Koch was not-too-jokingly referred to as "His Eminence." Koch's authority in the sphere of microbiology was so great that if he did not approve of a scientific finding, he simply dismissed it. Unfortunately, when he withheld his endorsement, it substantially contributed to delay in recognition and widespread knowledge about some important discoveries. He also was accused of tending to take credit for the work of other scientists in his laboratory.

Koch's most enduring legacy was the development of new and stringent laboratory methods that made it possible for any careful researcher to isolate microbes. His most original contribution was the elucidation of the

---

2. For TB statistics, see John H. Dingle, "The Ills of Man," in "Life and Death in Medicine," *Scientific American* 229, no. 3 (September 1973): 76–89; Raymond Obomsawin, "Immunization Graphs: Natural Infectious Disease Declines; Immunization Effectiveness; and Immunization Dangers," National Aboriginal Health Organization (December 2009), https://www.slideshare.net/MrChinLeeChan/immunization-graphs-by-r-obomsawin-2009.

germ theory of disease. It was revolutionary in his day to suggest that one germ was responsible for one disease. He made groundbreaking discoveries in many infectious diseases, including tuberculosis, anthrax, cholera, and malaria. By their own rules, the Nobel Foundation could not recognize Koch's life's work, so he was named for the 1905 prize "for his investigations and discoveries in relation to tuberculosis" despite the irony that he had promoted a spectacularly unsuccessful TB therapy.

The Nobel Committee had been considering Fritz Schaudinn, the discoverer of the syphilis bacterium, for the prize, but Koch's friends weighed in on his side. Schaudinn was ultimately passed over in favor of Robert Koch for the 1905 prize. Schaudinn died in 1906 and never got another shot at the Nobel.

6

The Mistaken Observer and Dr. Bacteria

The 1906 Nobel Prize was awarded jointly to Camillo Golgi and Santiago Ramón y Cajal for their contributions to knowledge of the structure of the nervous system. These two men independently studied *histology*, the branch of medicine that describes the microscopic structure of life-forms.

Golgi was an Austrian-born Italian who developed a method of staining tissues that differentiated their structural details. His procedure combined silver and potassium to create a black stain. This allowed minuscule cellular structures to show up in detail under the microscope for the first time. His first project in 1873 was the staining of brain tissue obtained from dead mental patients at the psychiatric hospital where he worked. The method is still called the *Golgi stain* to this day. It allowed the tracing of nerves throughout the body and brain. He thought various nerves that appeared to be crossing and recrossing other nerves were not separate. Golgi was convinced that they fused at these crossings, and all were part of the whole. He concluded that all nerves were part of one huge weblike interconnected network, which came to be known as the *reticular* (netlike) *theory*.

Within a few years, other histologists were getting experience with the Golgi stain. At the same time, the technology of microscopes was rapidly advancing.

Ramón y Cajal was a Spanish histologist who used the Golgi stain to conduct a much more detailed study of brain and nerves. He came to his own conclusions: nerves were part of a complex system that was not fused but composed of discretely separate cells, which he called the *neuron theory*. Ramón y Cajal's neuron theory was quickly shown to be the correct one, and Golgi's reticular theory was wrong. But his Golgi stain had made it possible for others to more carefully to study the nervous system, and it was for this that Golgi was awarded his portion of the prize.

By the time they were awarded the 1906 Nobel Prize, Golgi's reticular theory had been disproved for twenty-five years. Nevertheless, upon receiving his award, Golgi proceeded to deliver as his Nobel lecture a detailed description of the supposed reticular structure of the nervous system. Ramón y Cajal and other scientists in attendance were taken aback and embarrassed on behalf of Golgi, who was not the least bit embarrassed himself.

Santiago Ramón y Cajal was the son of a Spanish doctor who pushed his son to study medicine. However, the boy was more interested in photography, writing, and drawing. His father, a self-taught barber surgeon, tried to interest the boy in medicine by taking him to graveyards to find human remains to study. Ultimately, the young Ramón y Cajal found it easier to capitulate to his father's wishes and so ended up as a medical officer in the Cuban regiment of the Spanish Army in 1873.

Like so many others, Ramón y Cajal contracted malaria in Cuba. On the slow boat home, he saw the bodies of some of his comrades regularly slipped overboard after they had succumbed. Ramón y Cajal remained underweight and pale after the initial infection and struggled financially since he was not yet established in the medical profession. He found himself, at the age of twenty-four, not ever having kissed a woman, so despite his weakened state and poor financial prospects, he pursued the courtship of an orphaned girl who had been his pen pal during the war. In his autobiography he wrote that he did not know how to interpret the fact that he was not getting any kind of encouraging signs from her. Was it appropriate modesty, or had she gone cold on him? He put it to the test by suddenly planting a kiss on her lips. She recoiled in horror and called him a beast and a criminal. Having received his answer, he moved on.

Before he could take another shot at romance, Ramón y Cajal fell ill and suddenly coughed up blood. He had contracted tuberculosis, which was still potentially fatal in his day and had no known treatment. During his protracted stay at a sanatorium, he sunk into apathy and just waited for death to claim him. One day he woke up to the fact that he was not dying and jumped out of bed. Ramón y Cajal embraced his own new system of treatment that "consisted of doing everything contrary to the advice of the doctors" and made a complete recovery within weeks.

In the late 1870s, Ramón y Cajal's accurate histology skills earned him a position at the Faculty of Medicine at the University of Zaragoza, where he was able to continue peering into a microscope at beautifully stained neurons and then draw them in exacting detail. His more conventional colleagues scorned the microscope and considered his work foolish; they thought everything that needed to be known could be observed by examining the body with the naked eye. In medical schools of the day, Ramón y Cajal sadly observed that brand-new microscopes sat untouched by professors and students.

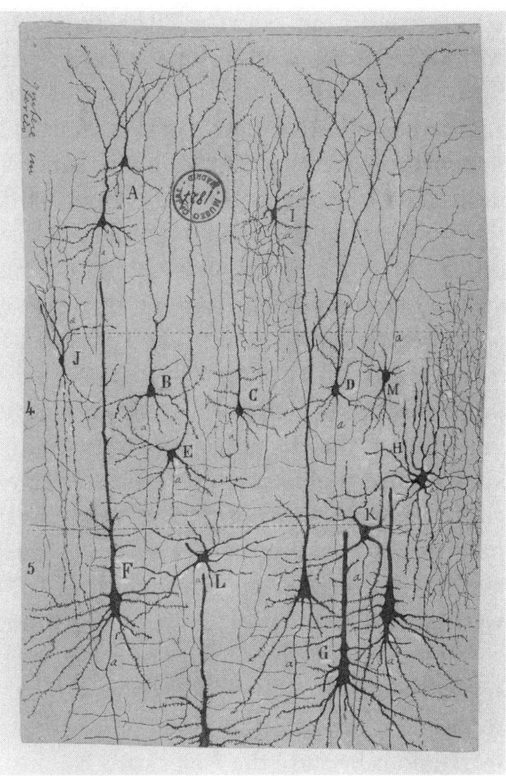

Among the first ever detailed drawings of brain neurons, by Ramón y Cajal. Courtesy of the Cajal Institute, "Legado Cajal", Consejo Superior de Investigaciones Científicas (CSIC).

Ramón y Cajal pursued marriage again now that he was healthy and had a good position. He married as soon as his new girlfriend showed interest, which brought condemnation from his friends and family. In those days, a professional life of scientific and academic pursuit was regarded by many as akin to joining a monastery. "Poor Ramón is lost for good! Farewell to study, science, and lofty ambitions!" they said. His father predicted his death in a short time, and his friends gave him up as having failed.

In 1885 and 1886, Ramón y Cajal took some vacation time to indulge his passion for writing. His series of science fiction tales, *Vacation Stories*, mostly had bacterial themes, and his pen name was Dr. Bacteria. "For a Secret Defense, a Secret Revenge" tells the story of how the infidelity of a young wife is detected by her older bacteriologist husband. He convinces himself of her liaison with his laboratory assistant from the printout of a seismographic device he had stealthily fitted to the laboratory couch. He infects the assistant with the TB bacterium and, when his wife comes down with it, he writes it all up in a scientific paper. The series also includes a horror story about a man who sees the world through microscopic vision with a magnification power of one thousand, a tale about a man who fails to convince the town that his house is not haunted but rather overrun by destructive bacteria, and a story about artificial insemination that follows the thoughts of the son of "a mother and a syringe." An early novel, which did not survive in print, was the story of a man on Jupiter within the blood vessels of a giant alien from which he describes "epic struggles between *leukocytes* [white blood cells] and parasites."

Meanwhile, Ramón y Cajal's career advanced with appointments to professorships in Barcelona and then Madrid. His photographs and drawings were revealing aspects of the structure of the brain and nervous system never before seen or even suspected. Yet they were underappreciated in Spain and not noticed at all outside the country. He began to publish papers in French that were subsequently picked up by the German medical journals. German recognition in 1889 finally put Ramón y Cajal's work on the international stage. This brought him into correspondence with researchers around the world, and he began to receive invitations to speak, honorary degrees, and visitors to his laboratories. In June 1899, he received an invitation to visit Clark University in Massachusetts. He was incredulous because the Americans had just defeated the Spaniards in the war not one year earlier.

The Spanish-American War began on a Monday in April 1898 and ended on a Friday in August of that same summer. It was instigated by American expansionists interested in establishing colonies in the tradition of European imperialism. The hostilities were strongly forwarded by Theodore Roosevelt, then assistant secretary of the United States Navy. The war occurred at a time in which the formerly great Spanish Empire was already withering; the short American campaign succeeded in dealing its deathblow. The Americans easily took Cuba, Puerto Rico, Hawaii, Guam, and even that namesake of Spanish royalty, the Philippines.

But the invitation from America had included a check to cover travel expenses, so Cajal took the trip. He had to maintain his composure while enduring various racial slurs such as this one from Clark University benefactor Stephen Salisbury: "I am delighted to entertain in my house a Spaniard who is endowed with some common sense!"[1]

1. Santiago Ramón y Cajal, *Recollections of My Life* (Cambridge, MA: MIT Press, 1989).

In recalling this in his autobiography, Ramón y Cajal thought that the remark communicated a deep insult. He also gives a hilarious description of a Fourth of July celebration:

> As a climax of ill fortune, the anniversary of Independence was being celebrated that day and a deafening clamour arose from the streets. Patriotic song, stentorian cheers, and the reports of rockets were heard, and, above all, shots, now singly and now in volleys. At the windows and on the roofs, I saw many people like madmen discharging rifles into the air. In the street, even the women were waving flags and shouting outrageously. Our traditional wrangles in the bull ring are child's play as compared with the frenzy and tumult of the American people on the famous Independence Day.[2]

In 1900, Ramón y Cajal fell ill again, this time attributing it to overwork and exhaustion. Admittedly a hypochondriac, he wrote: "I was attacked by neurasthenia, with palpitations, cardiac irregularities, insomnia, etc., with the resulting mental depression."[3]

Again he recovered without following any doctor's advice or taking quack remedies of the day. His cure was to move away from the city to the country.

It is not surprising that Ramón y Cajal used the term *neurasthenia*, which at that time referred to a mechanical weakness of the nerves. This was the era of an explosion in the number of discoveries in anatomy and physiology. For the first time ever, ills that had been considered purely sociological (contracted by poor people) or moral (such as venereal diseases) or mental were getting correctly attributed to physical diseases of the body, such as bacterial infection, parasitic infestation, or cancer. Conditions that had been vaguely attributed to ill humors or bad air were discovered to have their origins in actual physical pathologies. The exact anatomical descriptions of the nerves and brain cells by Ramón y Cajal and others were attached to grand hopes of discovering what it is that makes humans tick. It seemed only a matter of time before things like mechanical weakness of the nerves, for example, would be established to explain the various negative states of mind. No such actual weakness of nerve cells has ever been discovered.

When a telegram arrived from the Nobel Foundation in October 1906, Ramón y Cajal made nothing of it. In a few days, the media storm began and continued for months. This included congratulatory telegrams, letters

2. Ibid.
3. Ibid.

and messages from well-wishers, homage from professors and students, commemorative diplomas, honorary elections to scientific and literary societies, and even streets named after him. Other offers were less well received, such as "chocolates, cordials and other potions of doubtful hygienic value, marked with my surname" and "offers of profitable participations in risky or chimeral enterprises."[4]

Attendance at the Nobel ceremonies in Stockholm involved mingling with the other prizewinners in all categories. Ramón y Cajal was horrified and disgusted that American president Theodore Roosevelt had been awarded the Nobel Peace Prize that year. Roosevelt was a famous warmonger who was openly anti-Indian, anti-Chinese, and in fact generally anti-anyone who was not Anglo-Saxon. Historian Howard Zinn reports that Theodore Roosevelt wrote to a friend in 1897, "I should welcome almost any war, for I think this country needs one," and he carried around a list of six target nations on three continents. Zinn further quotes Teddy Roosevelt:

> "All the great masterful races have been fighting races," he claimed. To fellow Anglo-Saxons, he said, "It is wholly impossible to avoid conflicts with the weaker races," and added, "The most ultimately righteous of all wars is a war with savages."[5]

Cajal had dreams of bringing Spain up to speed with other developed nations in regard to educational standards in general and studies of science in particular. He worked to this end with other academics and political reformers until his death in 1934.

---

4. Ibid.
5. Howard Zinn, *A People's History of the United States* (New York: Harper, 2015).

7

The Obvious Suspects

The 1907 Nobel Prize was won by Charles Louis Alphonse Laveran for research into malaria. Even though his discoveries came before the later findings of Ronald Ross, Ross had gotten his prize five years earlier, in 1902.

Laveran had good reason to study malaria. His father was a military physician in Algeria during Laveran's youth. Malaria was *endemic* (always present) in Algeria, a French-occupied country in northern Africa on the Mediterranean. The disease struck people in swampy regions and was a chronic problem in many of the French imperial colonies.

When Laveran followed his father's career path and was assigned to Algeria in the 1870s, he observed that a great number of arriving French troops died even before they could muster up. He undertook to find the cause of malaria in a time when bacteria were newly being discovered as the cause of many diseases. Medical researchers of the day scoured the stagnant water in the malarial marshlands, sampled the humid air, and scrutinized the mucky soil to find the bacteria to blame. Twice it was announced that the malaria bacterium had been identified in the dirt, but others could not find the same bug. Laveran did all of the same investigations and likewise turned up nothing. Then he focused on the blood of malaria patients. Their red cells had dark granules, and in an effort to find out what caused that pigment, he unexpectedly saw a long-tailed parasite through his microscope. He then looked for it in others and reported finding the same organism in 148 out of 192 patients sampled.

Four years later, Italian researchers found some differently shaped parasites in the blood cells of malaria patients. With more study, it was determined that what they had observed was just a different stage in the life cycle of the same parasite originally described by Laveran.

The leading infectious disease researchers were skeptical because to them, all seemed to be bacterial. They rejected the existence of a new and previously unknown life-form. Their reluctance was a testament to how quickly they forgot that bacteria had been an extremely new concept but a few decades earlier. Bacterial researchers had faced the very same unwillingness of their elders to observe.

After a long time of skepticism, the famous bacteriologist Louis Pasteur was the first heavy hitter to acknowledge Laveran's parasite. After another decade, it finally became widely accepted that malaria is caused by a parasite.

Since the parasite was found in the blood cells but not in air, water, or soil, Laveran deduced that it must be carried by an intermediate organism. Mosquitoes were the obvious suspect since they were the most prevalent pest in areas of standing water. He published this hypothesis in 1884. This idea was not unique to Laveran. Patrick Manson, the renowned Scottish doctor of tropical medicine, had proved in 1877 that the African disease elephantiasis is transmitted by a parasite carried by mosquitoes and he had been the first to propose that malaria was also transmitted by parasites carried by mosquitoes. In 1883, the English-born American physician Albert Freeman Africanus King had suggested that to prevent malaria, Washington, DC, could be protected from mosquitoes by erecting a mesh fence as high as the Washington Monument to encircle the city.

Laveran held several military medical leadership posts before joining the Pasteur Institute as an esteemed contributor. He was eventually awarded membership as a commander in the Legion of Honor for his eminent contributions to French society. He donated half of his Nobel Prize winnings to the Pasteur Institute.

8

Immunity Wars

Ilya Mechnikov (also known as Elie Metchnikoff) and Paul Ehrlich were born nine years and nearly a thousand miles apart in the mid-nineteenth century, and they shared the 1908 Nobel Prize for their research into two entirely different ways the body fights infection.

The life of Russian-born Ilya Mechnikov was directly affected by infectious diseases of the day. His fiancée was so ill with tuberculosis that she had to be carried in a chair to the church for their wedding. When she died just a few years later, he was so distraught that he took an overdose of opium in an unsuccessful suicide attempt. Mechnikov remarried, and his second wife became ill with typhoid fever. She survived, but he was dejected again, and this time he tried a more creative suicidal gesture: he injected himself with a dose of serum from a disease called *relapsing fever*, which he had been studying. He survived that too.

The more usual way people get relapsing fever is from the bite of a tick or louse that transmits invading bacteria, although this mode of transmission was not known at the time of Mechnikov's experiments. Relapsing fever could be deadly, but many like Mechnikov did recover on their own. Today, relapsing fever is treated with antibiotics, yet one in a hundred still dies from it despite treatment.

Mechnikov's recovery marked a turning point in his studies: how does the body fight off an obvious bacterial invasion? He had the idea that organisms could neutralize invading bacteria by digesting them in the bloodstream. To test his idea, he stuck a thorn into a starfish larva. By the next day, he saw what he expected: the organism had engulfed and digested the foreign material. He named the process *phagocytosis* (from *phago*, "eat," + *cyte*, "cell"). Mechnikov looked for and found certain white cells in the bloodstream that devour bacteria and called these *phagocytes*. He was able to show that these cells had some kind of intelligence—they migrated purposefully toward a bacterium, engulfed it live, and then digested it. This was the discovery that earned him the 1908 Nobel Prize.

Mechnikov had worked at the university in Odessa, Russia, for twenty years, but the political climate surrounding the assassination of Alexander II began to change his life. In 1881, members of the anti-czarist group called People's Will threw a bomb that killed the czar, their third and finally successful attempt. This provoked a severe crackdown carried out by the repressive regime of the subsequent czar, Alexander III. Three times Mechnikov resigned his position in Odessa to protest the lack of academic freedom, and the last time he did not come back. He left Russia for good to work at the Pasteur Institute in Paris, where he remained for the rest of his life.

Mechnikov's studies at the Pasteur Institute were far more creative. He found that a specialized kind of phagocyte removes the coloring from hair. It was plain to see that white hair was a sure sign of aging, and so it occurred to him that maybe aging was not so normal after all. He proposed that aging occurred because the body's immune system was so frequently called into action that it eventually became indiscriminately activated. Aging, he thought, was the sum of effects of collateral damage coming from widespread immune system activity. He said that death was the final result of chronic inflammation caused by constantly fighting infections.

To support his theory, Mechnikov studied the age-related process of hardening of the arteries. He saw that the hard, artery-clogging material, plaque, was full of inflammation and often contained phagocytes in various stages of ingesting bacteria. He supposed the inflammation was due to chronic low-level bacterial infection. Indeed, it was shown much later that people who have chronic infections in any body part had a higher rate of death from heart attacks. This is especially true of infections in areas that had a rich blood supply such as gum disease, in which bacteria could easily spread into the bloodstream.

Mechnikov supposed that a lifetime of being under attack by microbes caused the phagocytes to be constantly active. The persistent action of phagocytes was generating side effects, especially damage to bones, kidneys, the brain, blood vessels, and the heart; this culminated in all of the classic signs of aging: brittle bones, shrunken kidneys, dementia, and hardening of the arteries. Thus, he concluded that aging could be prevented, and he doubted even that death was natural.

Meanwhile, he was stumped by the fact that humans carried around trillions of harmful bacteria in their own large intestines. He was fascinated by the case of a woman who had her colon surgically removed but lived well without it. He concluded that the large bowel was not necessary to life but, rather, was only a reservoir of dangerous bacteria. Mechnikov thought it would be a good idea for everyone to have his or her colon removed. He lamented it was too bad that society was not enlightened enough to submit to mass operations to live longer. Realizing that would never change, he worked to find a way to lessen the effect of the major load of colon bacteria. He and his colleagues at the Pasteur Institute found that the bacteria that caused milk to ferment created a weak acid, lactic acid. When people ate yogurt, lactic acid produced by the acidophilus bacteria caused a die-off of the bad bacteria in their colons.

Mechnikov was the world's first promoter of what today are called probiotics. Although he did not specifically recommend it, he would probably approve of today's chewable probiotics, which have been proved to decrease bad bacteria in the mouth.

Besides yogurt with live acidophilus cultures, Mechnikov's antiaging recommendations included not eating any meat since it tends to rot in the gut. He advised strict avoidance of all raw foods and boiling food whenever possible to kill any dangerous soil organisms. He urged public education measures to end alcoholism and syphilis, two conditions that had already been proved to promote premature heart disease.

Mechnikov's most intriguing antiaging idea was revolutionary. He saw that the blood from organs of one animal caused deadly blood clotting when injected into another species. However, if the cells of one animal's organs were injected into another species in very small amounts and this procedure was repeated over several weeks, then the receiving animal tolerated it without blood clotting. After a few weeks, the serum of the second animal was drawn off and, when injected into a third animal of the original species, worked like medicine to treat diseases of the organs. He described these steps for use in humans: First, find some fresh human organs, preferably immediately after death. Mince up the organ tissue and inject a horse with small doses of the diluted mixture. After several weeks of injections, draw the serum from the horse and use it to treat humans who have organ disease such as kidney failure, heart failure, or liver failure. His proposed treatments never made it into mainstream medicine, but we can see traces of this experiment in modern-day stem cell treatments.

Paul Ehrlich owned the second half of the 1908 Nobel Prize's story. Like Mechnikov, he was personally touched by infectious disease. He came down with tuberculosis (TB), which some historians thought he may have caught while working with TB bacteria in various German research laboratories. He spent a couple of years in Egypt in recovery. The theory then was that warm, dry climates assisted in TB recovery.

Along with other prominent researchers of the day, Ehrlich initially rejected Mechnikov's idea of phagocytosis because it clashed with the so-called serum theory of immunity. It had already been shown that when persons survived an infectious disease, their blood had something in it that protected them from getting that disease again. Ehrlich was determined to find out about the nature of the mysterious serum factor and set forth an idea that the body's infection-fighting cells were ingenious custom chemical factories. Contact between invading bacteria and the surface of a white blood cell caused the smart cell to start making a chemical arm. That arm was custom-made to lock onto the bacteria very specifically, rather like a custom-shaped parking spot. There the bacteria were immobilized and rendered harmless. Eventually, other researchers took up Ehrlich's theory and were able to demonstrate the existence of what we today call antibodies.

Ehrlich thought these chemical sidearms popped up all over the white blood cell surface, ready to waylay any other bacteria of the same type; in addition, antibody sidearms were sent out into the bloodstream as free agents to seek and lock up floating bacteria. Furthermore, he detected some other biochemicals in the blood that aided the destruction of bacteria and called them *complement* because they complemented the action of the antibodies. He hypothesized that complement was as specific as the antibodies—in other words, one kind of complement was produced by the body on demand when there was an infection, and it was capable of attacking only one kind of bacteria. Ehrlich received his portion of the Nobel Prize for his discovery of antibodies and his theory of complement.

Ehrlich and Mechnikov exchanged harsh criticisms of each other's ideas about immunity. It was eventually shown that depending on the infection, the body produces antibodies or digests bacteria with phagocytes, or does both.

Ehrlich's later work was tremendously beneficial to society, but it brought him much grief. For example, despite playing a huge role in the development of the vaccine for diphtheria, he was squeezed out of most of the profit from it by his coworker Emil von Behring, the first Nobel Prize winner. Ehrlich remained bitter that he was cheated out of money but was more disgruntled that he was passed over for the Nobel Prize in 1901. The Nobel Committee had received only a single nomination for Ehrlich in 1901, and it was submitted by Mechnikov. By the time he finally won, Ehrlich had been nominated seventy-six times.

Ehrlich fared better financially a decade later when he allowed Hoechst to use the formula for his revolutionary new antisyphilis drug, Salvarsan. But he was attacked in the papers for growing rich from work he did in a state-funded laboratory, as well as for supposedly testing the drug on unsuspecting German prostitutes. It is difficult to be sympathetic with Ehrlich since he excluded his codiscoverer, Sahachiro Hata, from the Hoechst deal.

Discovery of Salvarsan came out of a unique system Ehrlich devised that is still the way drugs are developed today. He systematically tested the classes of bacteria against categories of possible drugs and cataloged all reactions. For example, if small, rounded bacteria reacted to mercury by shrinking, then various mercury compounds should be invented and tested on all kinds of small, round bacteria. If rod-shaped bacteria curled up and died when exposed to acid, then various acid formulas should be concocted and tested on all rod-shaped kinds of bacteria. He hoped that this testing system would lead to what he called a "magic bullet" for each disease. Just as he knew the body manufactures custom antibodies specific to particular bacteria, he hoped scientists could make chemical compounds that attacked specific germs. Ehrlich coined the word *chemotherapy*. Today chemotherapy is a word we associate only with cancer therapy, but it is a nonspecific term that simply means chemical therapy.

By the early 1900s, arsenic had been used off and on for centuries to treat syphilis, but it also poisoned many patients. Working in Ehrlich's lab, Sahachiro Hata applied the system to testing various less-toxic arsenic-containing compounds. The sixth compound in the sixth category tested was a winner: it was poisonous to the bacteria but not excessively toxic when given to lab animals. This man-made chemical was 36 percent arsenic and was initially called simply "606." (Some reports say it was the 606th compound tested.) Ehrlich renamed it Salvarsan, and it was the first *synthetic* drug in the world, meaning it was entirely man-made and did not exist in nature.

Per Ehrlich's new terminology, using his synthetic drug became known as the first chemotherapy. But some people were furious with the discovery.

Salvarsan was accused of being a dangerous drug pushed on patients as if it were a miracle cure. In fact, Salvarsan routinely caused nausea and vomiting and could also cause rashes, liver damage, and a debilitating, painful inflammation of the nerves affecting the arms and legs, called polyneuritis. These were all signs of arsenic poisoning. Another vein of social attack on Salvarsan was claiming treatment for a venereal disease would only promote immoral behavior, and some thought that curing syphilis interrupted the natural cycle of crime and punishment. Even the name of it was attacked—606 contained numbers associated with the devil. Racism was thrown into the mix, as the Jewish Ehrlich and the Japanese Hata were accused of unspecified ulterior motives.

These attacks were proclaimed by an unlikely trio: Dr. Breuw, a German nationalist physician; Karl Wassman, a weird pamphleteer who walked the streets in a monk-like robe complete with rope belt; and the Lutheran clergy. Finally, supporters of Ehrlich and Hata brought a libel suit against pamphleteer Wassman. This brought Ehrlich to the witness stand to defend his clinical testing of Salvarsan. A *New York Times* article dated June 9, 1914, quoted Ehrlich's testimony under cross-examination in which he admitted that there were some reports of patients dying or becoming paralyzed after taking Salvarsan.

Eventually, the drug formula was tweaked to be less toxic, but that also made it less effective against syphilis. Ultimately Salvarsan was replaced by penicillin. Sahachiro Hata was nominated three times for the Nobel Prize for his considerable role in discovering the syphilis remedy but never won it. Another Japanese researcher, Hideyo Noguchi, was never recognized by the Nobel Committee for his identification of the causative agent of syphilis—the bacterium *Treponema pallidum*—in the brain tissues of patients suffering from partial paralysis due to the inflammation caused by the bacterium. Decades later, the anti-Semitic arguments against Ehrlich were revived. In 1938, the Nazis pulled down street signs along Paul-Ehrlich-Strasse (street) in Frankfurt.

Ehrlich's final research focused on solving the riddle of what the body did to resist cancer. He had a theory that tumors could be starved out, and modern cancer research continues to study this possibility. Ehrlich died in 1915, and Mechnikov died the following year, his colon still intact.

9

Accidental Harm, Chocolate, and the Nobel Prize

Emil Theodor Kocher won the 1909 Nobel Prize for the work he did on the thyroid gland. He was a Swiss physician born in 1841.

By the time Kocher had made it through his medical training, surgery was still a dirty mess and many patients died. Kocher embraced the ideas put forward by the English scientist Joseph Lister in 1867, who was the first to suggest the use of chemicals to kill germs at the time of surgery and in treating wounds. The practice was called *antisepsis* (from *anti*, "against," + *sepsis*, "infection"). Listerine was named in his honor in 1879, which seemed to please him. One wonders if Joseph Lister would have approved of several bacteria and a slime mold that were also named for him posthumously.

The use of antiseptics decreased the chance of developing an infection after surgery, but there were still too many postoperative deaths. Eventually, both Lister and Kocher independently combined the use of antiseptics with the practice called *asepsis* (from *a*, "without," + *sepsis*, "infection"). In aseptic surgery, physicians make sure to protect the patient from any bacteria the surgeons are carrying on their skin or through their breath. Surgeons began to scrub their hands between patients; wear gowns, gloves, and cloth face masks; and use only sterilized surgical instruments, stitching thread, and sponges.

Kocher was convinced that a bloody surgery was more likely to result in infection, so he took extra care in his operations to keep bleeding to an absolute minimum and meticulously cleared away any bit of blood. This slowed down his operations to a pace that his students called "Kocher time," but it worked: his patients had very low death rates.

Kocher developed specialized surgical techniques for operating on the thyroid gland. This gland is in the front of the neck, just below the Adam's apple. In Kocher's time, most doctors presumed the thyroid gland to be a useless vestigial organ, something left over from our animal predecessors, like the appendix. When a person developed enlargement of the gland (a *goiter*), it was often ignored because surgery on it was considered very risky. But with the advent of aseptic surgery, thyroid removal became a safer procedure. Kocher perfected the delicate *thyroidectomy* operation, which has to be carried out with care so as not to damage the windpipe or the nerves that control the ability to make speech. He did thousands of these operations, and his hospital became a world center for doctors to train and the rich or famous to get treated, including, for instance, Lenin's wife, who was operated on in 1913.

Kocher's patients had a very low death rate, but others brought to his attention that many of his patients did not do so well months and years down the road. He tracked down hundreds of former thyroidectomy patients. Some of them were just fine; these were the ones on whom he had done *subtotal thyroidectomy*, in which not all of the thyroid gland was removed. Some of these patients even regrew goiters but still were feeling fine. Other patients were doing badly—they had gained weight, were retaining water, and developed coarse features, brittle hair, and mental retardation. These were the ones from whom Kocher had removed the entire gland (*total thyroidectomy*). Months or years after surgery, they had turned into cretins. *Cretinism* was a condition the Swiss were familiar with because they had seen many children in certain villages born with it. Here is a description from 1855:

> I see a head of unusual form and size, a squat and bloated figure, a stupid look, bleared, hollow and heavy eyes, thick projecting eyelids, and a flat nose. His face is of a leaden hue, his skin dirty, flabby, covered with tetters [itchy skin pustules], and his thick tongue hangs down over his moist livid lips. His mouth, always open and full of saliva, shows teeth going to decay. His chest is narrow, his back curved, his breath asthmatic, his limbs short, misshapen, without power. The knees are thick and inclined inward, the feet flat. The large head drops listlessly on the breast; the abdomen is like a bag. Generally the cretin is deaf and dumb, or able to utter only a hoarse cry. He is indifferent to heat, cold, blows, and even the most revolting odours.[1]

Cretins have been described since ancient times, and there are reports of Alpine mountaineers coming upon whole villages of them. The cause was unknown. When Kocher realized that his technically perfect operations were causing cretinism, he expressed his horror: "I have doomed people

1. *Encyclopædia Britannica*, 11 ed. (1911), s.v. "cretinism."

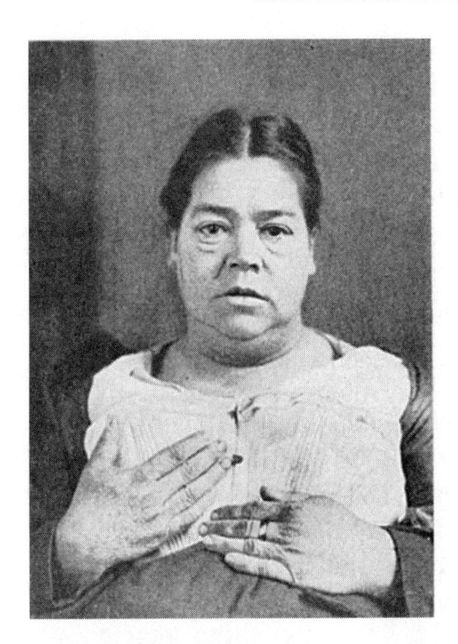

Hypothyroid patient before and after treatment with thyroid hormone

with goiter, otherwise healthy to a vegetative existence. Many of them I have turned to cretins, saved for a life not worthy living."[2]

Not everyone thought the thyroid was a nonfunctioning body part. A physician wrote in 1838 of his suspicion that the thyroid gland secretes necessary substances, and before Kocher, other surgeons had reported cretinism in their thyroidectomy patients. It was the sheer large numbers of thyroid operations done by Kocher that brought to light that the thyroid gland was making something biologically active that prevented cretinism. Kocher prayed, and he wrote essays on the ethics of the physician who is supposed to "do no harm." He resolved to never again do a total thyroidectomy, instead leaving a little bit of thyroid tissue to prevent cretinism.

Other researchers analyzed thyroid tissue and found it was extraordinarily high in the element iodine. Eventually, it was found that regions with a lot of cretinism (in newborns and acquired later in life) had a lack of iodine in the soil and, thus, in the food. There were several ways to develop cretinism, including being born to a mother who had iodine deficiency during pregnancy, growing up in an iodine-deficient area, or having your

2. Saurav Sarkar et al., "A Review on the History of 'Thyroid Surgery,'" *Indian Journal of Surgery* 78, no. 1 (February 2016): 32–36 (quoting Emil Theodor Kocher, "Über Kropfextirpation und ihre Folgen," *Archiv für Klinische Chirurgie* 29 [1883]: 254–337).

entire thyroid gland surgically removed. It turned out that iodine supplementation during pregnancy prevented newborns from being cretins; in children and adults, iodine also prevented a goiter from forming, and it caused established goiters to shrink or go away completely. But iodine did not help people who already had their thyroid glands surgically removed. It was deduced that something in the thyroid gland was using iodine to make a biologically active substance.

Kocher experimented with transplanting thyroid glands from animals into humans, pioneering the practice of organ transplant. It seemed to work. Other treatments that worked included injections with a solution of ground-up sheep's thyroid gland and giving animal thyroid tablets by mouth. It was not until 1926 that the structure of thyroid hormones was finally determined (by Edward Kendall, who won the Nobel Prize in 1950). Iodine was found to be a necessary part of the treatment. Iodine is an integral part of the structure of thyroid hormones; if there is not enough iodine in one's food, then thyroid hormones do not get manufactured by the gland.

In his Nobel speech, Kocher omitted giving any credit to some prominent thyroid researchers who assisted and complemented his work. Alfred Nobel's will specified that the Nobel Prize should be awarded to individuals, not teams, and that no more than three people could share one year's prize—creating an issue of due credit to assistant researchers that is continually present in the Nobel Prize awards and acceptance speeches.

Kocher is remembered far beyond his perfection of safe surgical techniques. A number of surgical procedures are named after him, as well as some abnormal physical examination signs and many different surgical instruments.

The 1910 Nobel Prize was awarded to a German named Ludwig Karl Martin Leonhard Albrecht Kossel, born in 1853. Kossel determined the chemical composition of the substance in the nucleus of the cell (*nucleic acid*). He essentially discovered what would later be identified as the genetic information found in all living cells, which provides the molecular structure necessary in the formation of stable DNA and RNA molecules. Since then, there have been fourteen more prizes awarded for discoveries involving genetic research.

While a professor of physiology and director of the Physiological Institute at the University of Marburg from 1895 to 1901, Kossel was the first to isolate *theophylline*, a therapeutic drug found naturally in tea and cocoa beans. Theophylline is in the class of medications called *bronchodilators*. It relaxes the muscles of the airway passages in the lungs. Theophylline quickly became a mainstay of treatment for asthma, bronchitis, and emphysema. In 1901, Kossel was named to a similar post at Heidelberg University and became director of the Heidelberg Institute for Protein Investigation. His research into the chemical composition of proteins paved the way for the later discovery of the *polypeptide* nature of protein molecules (a long string of linked amino acids).

Kossel was from a country with relatively high chocolate consumption. A recent study showed that countries with high chocolate consumption also produce the most Nobel laureates per capita. Top-ranking Nobel/chocolate correlations were found in Switzerland, Sweden, Denmark, Austria, Norway, Germany, and Ireland. It is not known if Kossel personally liked chocolate.

The 1911 prize was given to Swedish-born Allvar Gullstrand for working out the mathematical basis of the vision-focusing problem called *astigmatism*. In astigmatism, an irregularly shaped cornea or lens prevents light from focusing properly on the retina (the light-sensitive surface at the back of the eye). As a result, vision is blurred at any distance. Although Gullstrand had no formal training in physics, his mathematical work in optics was definitely in the realm of physics. He was nominated for a Nobel Prize in both medicine and physics but rejected the physics prize and instead accepted the prize in medicine.

Gullstrand's other claim to fame was for the role he played in preventing Albert Einstein from receiving a Nobel Prize in Physics for his famous theory of relativity. Appointed to the Nobel Committee for Physics, Gullstrand evaluated the nominations of candidates for the Nobel Prize. Gullstrand and his French colleague, Paul Painlevé, worked out an alternative and, they thought, more accurate solution to one of Einstein's formulas. They invited Einstein to a public debate so he could have the opportunity to defend his calculations. Einstein at first avoided the invitation to an open debate on the issue. When he finally appeared for the debate, the focus was on the use of a mathematical formula devised to calculate the velocity of a raindrop as it falls toward a black hole in space. Einstein admitted he was baffled by the mathematical terms used in the Gullstrand-Painlevé equations. On the other hand, Einstein's explanation, using his famous theory of relativity, was not understood by most mathematicians or physicists. In fact, Gullstrand was one of the few who did understand it, but he thought the equation was outright erroneous.

Einstein was nominated for a Nobel Prize ten times between 1910 and 1922, and strong lobbying for and against went on behind the scenes every year. Gullstrand's repeated objections to awarding Einstein the physics prize were that the theory of relativity was strictly theoretical and had never been proved, nor did it fulfill the requirement of contributing to the good of humanity. Gullstrand wrote strong position papers against awarding the prize to Einstein. Finally, the committee voted to award Einstein the 1921 Nobel Prize in Physics for his work in photoelectricity, an area that had already yielded practical value. Einstein was sent a letter specifically stating that he was not awarded the prize for his relativity theory. The photoelectric principles Einstein worked out are in use today in limiting static electricity on spacecraft and in the design of night-vision devices.

10

The Perpetual Chicken

Alexis Carrel won the 1912 Nobel Prize for his techniques of sewing blood vessels and facilitating organ transplantation. Carrel was born in France in 1873. His widowed mother, who was a professional embroiderer, provided his early education at home.

Carrel's interest in surgery began in childhood with experiments on small animals. The legend goes that he was distressed by how, in 1894, the French president Marie François Sadi Carnot bled to death two days after an assassin's knife wound. (Surgical techniques of the day could not reliably patch up injured blood vessels.) Carrel practiced vascular surgery on animals, finding that he had to solve several problems: the stitches had to be tiny enough to sew thin vessels without creating fresh bleeding at every needle hole; the surgery had to be done in a way that did not cause the blood to clot on irregular surgical edges; the rejoined vessel had to be wide open for blood flow (and not pinched or narrowed at the repair site); and the surgery must not introduce infection into the patient.

As a young man, Carrel studied with a master embroiderer in Lyon, France, practicing with fine thread and delicate instruments. Throughout his medical training, he perfected his technique on animals. *Carrel's seam* is a method of turning back the vessel ends to make cuffs, then joining the cuffed ends with tiny stitches.

Carrel became a doctor of medicine in 1900. According to contemporary accounts, he was not well liked at the medical center because of his abrasive, self-important personality and unconventional ideas. He was scorned after calling for scientific investigation of the faith-based miracle healings that took place at the Catholic shrine of Lourdes, one of which he had witnessed. As his colleagues were mostly devout Catholics, they objected to Carrel's insistence that some scientific principle must be found to underlie what was generally understood to be a spiritual occurrence.

Eventually, Carrel failed twice to pass the exam to become a full faculty member at Lyon, prompting him to announce plans to emigrate from France and become a cattle rancher in Canada. The cattle plan never materialized. Instead, he took a position at the University of Chicago, where he experimented with vessel surgery and organ transplantation. There he earned the animosity of his colleague and coworker Charles Claude Guthrie. When Carrel submitted scientific papers on their research, he left off Guthrie's name.

John D. Rockefeller saw something he liked in Carrel and offered him a position at the newly formed Rockefeller Institute in 1906. By the time he was awarded the Nobel Prize in 1912, Carrel's suturing techniques were in use all over the world.

Carrel's research interests broadened to include investigating how to make organ transplantation practical. He took heart tissue from a chicken embryo and stirred it into a flask of warm nutrient broth. His laboratory assistants were able to keep a culture of the tissue alive and cells dividing for over thirty years, far exceeding the usual life span of a chicken and even outliving the doctor. The scientific record on these supposedly immortal cells is very sketchy, and it was rumored that whenever the growth petered out, Carrel would just order additional fresh embryo cells to be tossed into the flask. Nevertheless, the experiment prompted sensational headlines that warned of a science-fiction age populated by man-made creatures. By the time Carrel was a prominent figure in the Rockefeller laboratories, it was said that he mostly wrote scientific papers based on the actual work in the laboratories done by others.

Carrel had some success in transplanting organs in animals but had a problem preserving donor organs until they were ready for transplant. This was solved by his friend Charles Lindbergh, the famous pilot who was awarded the Medal of Honor for flying his single-engine plane nonstop from New York to Paris in 1927. Lindbergh initially sought out Carrel to see if his sister-in-law's heart, damaged by rheumatic fever, could be repaired. They became close friends through working together at the Rockefeller Institute and sharing personal, political, and social views. When Lindbergh saw the crudeness of Carrel's machinery, he offered to build new equipment for the scientist. Eventually, they built the first *perfusion pump*, a device that kept donor organs alive by pumping them with blood until they could be transplanted into the recipient. A similar device kept the body pumped with blood during transplant operations. Lindbergh devised a respiratory chamber that kept the organ infused with oxygen during temporary storage. These inventions would make it possible to do the first open-heart surgery using a heart-lung bypass machine, which was accomplished by Dr. John Gibbon in 1953.

In 1935, Carrel wrote a book for the lay public called *Man, the Unknown*. The 1936 version of this book included a preface that praised eugenics, scrubbed from later versions. In *Man, the Unknown*, Carrel advocated authoritarian government, a limited role for women outside the home, and, above all, the application of scientific measures to "moral and intellectual decay." By this last comment, he meant the use of *eugenics*, the techniques of supposedly improving a human population by controlled breeding to increase the occurrence of desirable heritable characteristics. Carrel said:

> Eugenics is indispensable for the perpetuation of the strong. A great race must propagate its best elements. . . . It is known that children born in families of superior people are more likely to be a superior type than those born in an inferior family.[1]

Carrel ranked psychology as the "supreme science," which should be emphasized above all of the other (inferior) sciences such as physiology, anatomy, mechanics, chemistry, physics, and mathematics. He envisioned the creation of a "high council" of a few men who would use psychology to ensure the survival of the minds and souls of a "great race."

In *Man, the Unknown*, Carrel recommends:

> Those who have murdered, robbed while armed with automatic pistol or machine gun, kidnapped children, despoiled the poor of their savings, misled the public in important matters, should be humanely and economically disposed of in small euthanasic [*sic*] institutions supplied with proper gases. A similar treatment could be advantageously applied to the insane, guilty of criminal acts.[2]

Carrel had an idea that since a fetus carries something of the man, it had a balancing effect on the nervous tendencies of the woman:

> Women who have no children are not so well balanced and become more nervous than the others. In short, the presence of the fetus, whose tissues greatly differ from hers because they are young and are, in part, those of her husband, acts profoundly on the woman.[3]

1. Alexis Carrel, *Man, the Unknown* (New York: Harper & Brothers, 1935); view the free 1939 version at http://www.soilandhealth.org/03sov/0303critic/030310carrel/Carrell-toc.htm.
2. Ibid.
3. Ibid.

Carrel was forced to retire from the Rockefeller Institute in 1939 in accordance with its policy for mandatory retirement at age sixty-five. He returned to France and led the Foundation for the Study of Human Problems, staffed by three hundred psychologists, doctors, and statisticians. Its stated goal was to improve the welfare of France's children, but it appeared to be an institution that promoted the application of the principles of eugenics to social issues. The foundation was generously financed by the Vichy government, a puppet administration set up by the occupying Nazis.

For this work, Carrel was labeled as a collaborator with the Germans, which was an obvious conclusion given the ideas he expressed in *Man, the Unknown* and his close collaboration with Nazi sympathizer Charles Lindbergh. Carrel was forced to resign in 1944 when the Americans liberated Europe from Nazism. Alexis Carrel died on November 5, 1944, in France, thus avoiding the Nuremberg trials for war crimes against humanity.

11

Another Eugenicist

Charles Richet earned the Nobel Prize in 1913 for his observations of the body's response to foreign substances. Richet was born in France in 1850. He was trained as a conventional medical researcher but also strayed far afield into metaphysics, extrasensory perception (ESP), and hypnotism.

Richet made an accidental discovery one day while attempting to vaccinate laboratory dogs against jellyfish poison. He saw that when an animal is exposed to a small amount of poison, it might have no reaction or only mild discomfort. When given the same substance a few weeks later and in an extremely diluted dose, the animal would respond violently with an immediate itchy, welt-like rash, vomiting, bloody diarrhea, convulsions, and even death.

Richet termed this phenomenon *anaphylaxis* (from *ana*, "opposite," + *phylaxis*, "protection"). He reasoned that the goal of the body was to protect itself from foreign substances. But sometimes these reactions were excessive, and instead of protecting the body, especially vigorous reactions caused harm and possible death.

Anaphylaxis explains why a person may be given a dose of penicillin or general anesthetic and experience no adverse effects but have a deadly allergic reaction upon their very next dose. Richet's experiments showed the first exposure caused some chemical change in the blood that was activated only when the second exposure occurred. This work paved the way for understanding allergies, sensitivities, and intolerance to drugs or particular foods. His demonstrations are currently being cited as the likely explanation for chronic autoimmune illnesses that follow vaccinations.

Richet had a lifelong interest in hypnosis, experimenting on his schoolmates when he was a teenager. His medical training exposed him to the teachings of Dr. Jean-Martin Charcot, the director of the famous Salpêtrière Hospital in Paris, who promoted the idea that all mental distress had a physical basis. Charcot formed a neurology department to treat thousands of women determined to have *hysteria* (from the Greek *hystera*, "womb"), which he taught was a hereditary disorder characterized by a weak nervous system peculiar to patients with wombs. Charcot promoted hypnotism as a supposedly scientific cure. In an agreeable subject, the hypnosis caused an outburst of hysteria and thereby (supposedly) allowed the patient to "work through" her madness. The abundant surviving medical literature by Charcot and his followers does not report many improvements.

Richet was also a contemporary of Wilhelm Wundt and William James. Along with Charcot, these leading medical men of the day supported the idea that thinking and behavior had entirely physical foundations in genetics, physiology, anatomy, and biology. Wundt ceaselessly advanced the idea that humans were but animals, although genetically advanced animals. James studied spiritualism by experimenting on himself with knockout gas (chloral hydrate), nitrous oxide, and peyote. Through his hallucinogenic experiences, he somehow concluded that emotions were dictated by bodily functions.

Richet's study and participation in psychic experiences were driven by his interest in how the brain itself was formed from the combination of genetics and life experience to become "the mind." He proposed that a great network of nerves determined human thought and behavior. All of these theories mirrored the growing antireligionist sentiment of the day: everything was material, even the spirit.

Richet enthusiastically followed the career of an apparently telepathic Spaniard, Joaquín María Argamasilla, who had "X-ray eyes" that allowed him to read through metal although blindfolded. Harry Houdini exposed the alleged psychic to be a fraud during Argamasilla's American show tour. Richet was entranced by the medium Eva Carrière, who eventually admitted to hiring her Arab coachman to pose as a cloaked three-hundred-year-old Hindu who sneaked into the séance room through a trapdoor. Richet continued to invite the psychic Eusapia Palladino to hold séances at his home even after the Society for Psychical Research had exposed her trickery in moving furniture and touching participants during a supposed meeting with the dead.

Richet was as prominent a figure in what he called "metapsychics" as he was in medicine. He was known to speak to a conventional academic assembly of scientists in the morning and follow it up that same afternoon with a public presentation on metapsychics to a packed hall of mediums and séance enthusiasts. He served as president of the Society for Psychical Research in 1905, a post previously held by William James. The nonprofit UK-based society still exists today.

Dabbling in creative writing under the pen name Charles Epheyre, Richet wrote the novella *Soeur Marthe*, in which a young doctor hypnotizes a neurotic woman into a catatonic state. Also as Epheyre, he wrote the play *Circé* based on the Greek mythical goddess who turned men into beasts. In 1890, he wrote a science fiction tale set forty-five years in the future, "Professor Bakermann's Microbe," in which an evil biologist creates a strain of especially virulent superbacteria in 1935. It spreads disease in minutes, kills in an hour, and has no known antidote. Bakermann was not upset when his nagging wife was the first to succumb.

Cover of Charles Richet's book of social commentary, *L'Homme Stupide* (*Idiot Man*), published in 1919

In another unique sideline, Richet was a collaborator with fellow Frenchmen Louis and Jacques Breguet in the design of a sort of rudimentary helicopter. It is reported that in the summer of 1907, Gyroplane No. 1 and its pilot rose vertically about two feet and stayed aloft for one minute while four people held it steady.

Richet had served as a medical corpsman in the Franco-Prussian War in 1870 and thereafter became a vocal antiwar advocate. He based his rationale on the tremendous financial cost of war rather than on any humanitarian grounds. He was a member of the International Peace Society (Société de la Paix), followed by the League of Arbitration between the Nations (Société française pour l'arbitrage entre nations), eventually leading the league as its president and participating in international peace conferences.

Richet's insistence on international peace may not have been all it appeared on the surface. It turned out that he was just as vocal in his racist beliefs, strongly bolstered by his philosophy that humans were merely complex biological creatures devoid of an inherent ethical quality or spiritual aspect. He did not think highly of humankind, and in 1925, he wrote a book called *L'homme stupide*; the English translation is titled *Idiot Man*. He listed all of humans' ineptitudes including drinking, taking drugs, smoking, catching venereal diseases, getting tattooed, making war, letting mosquitoes live, fearing death and free trade, destroying the great forests, and modern humans' failing to embrace the international language of Esperanto. He said, "Yes, in my inmost being, I am humiliated because I belong to this vile animal species, the most foolish of all created things."[1]

Richet was president of the French Eugenic Society from 1920 to 1926. The society was founded in 1912, and until Richet assumed leadership in 1920, it had promoted so-called positive eugenics. It supported hygiene measures, healthy pregnancies, breast-feeding infants, and active child rearing to strengthen the French population in numbers and vitality. But in 1920, the society began a campaign of negative eugenics that lasted through the pro-Nazi Vichy government years of the 1940s. The society lobbied for forced premarital exams to weed out persons deemed unfit for procreation. Meanwhile, Richet was calling for an end to the arms trade and at the same time admiring the efficiency, discipline, and weapons superiority of the German Army.

Richet died in 1935. Ironically, his son, grandson, and a grandnephew were put in concentration camps in World War II, and another grandson and his daughter-in-law were held as German prisoners of war, all for presumed Jewish connections.

1. Charles R. Richet, *Idiot Man; or The Follies of Mankind* (*L'homme stupide*) (London: T. Werner Laurie, 1925).

12

Balance

In 1914, the Nobel Committee could not find a candidate worthy of the prize in medicine, and so none was awarded. The committee may have been distracted by the outbreak of the Great War, also called the War to End All Wars, but by the next year it had dug a little deeper. In 1915, committee members determined that the work done by Róbert Bárány in 1906 was deserving of their recognition. Bárány was named the winner in 1915, but it was called the Nobel Prize "of 1914." In fact, Bárány did not travel to Stockholm to receive the prize until 1916.

Róbert Bárány was born in 1876 near Vienna, in Austria-Hungary (today's Austria). As a child, he suffered from a tuberculosis infection of the bones, which left him with a limp for the rest of his life but also sparked his interest in medicine. Bárány became a doctor in 1900 and then studied surgery as well as neurology and psychiatry.

Bárány studied under Emil Kraepelin, who was famous for naming every objectionable behavior as if it were a physical mental illness. Mental retardation, criminality, epilepsy, homelessness, and prostitution were all considered by Kraepelin to be psychiatric disorders. However, he himself admitted that doctors could not reliably distinguish where normality ended and pathological mental states began. Unlike his contemporary Freud, Kraepelin forwarded the idea that mental distress had its basis in abnormal brain anatomy. He was a eugenicist who believed that all Jews suffered a genetic defect in the brain.

Bárány was also a student of Freud but scorned some of Freud's theories, including his famous dream interpretations. Freud taught that all dreams reflected some unconscious desire. Bárány reported having had a dream that did not involve any desire. Freud concluded, "That is very simple. You had the desire to contradict me." Ironically, after her father's death, Bárány's only daughter, Ingrid, had a long and successful career as a Freudian psychoanalyst in America.

Bárány eventually settled on the specialty of the inner ear, firmly establishing that the ear had important functions besides hearing. He made careful observations of the ear's mechanisms for balance and orientation in space. He saw that when the ear canal was irrigated with warm or cold water, it caused the eyes to turn to one side or other. The direction of eye movement depended on water temperature. When warm water was instilled into the ear, the eyes rolled to the opposite side. It was exactly the reverse when ice water was instilled: the eyes rolled to the same side as the irrigation. Bárány figured these reactions to water temperature reflected the normal ear mechanisms that allow us to maintain balance through a combination of mechanical and neurological events. He hypothesized that the warmth in the ear canal was transmitted to the inner ear, and the heat caused the level of lymph fluid inside the ear to rise ever so slightly. In turn the heat increased the rate of firing of nerve cells on that same side, nerve cells that were connected to the brain. To the brain, this mimics what would happen when the head is turned in that direction. In contrast, Bárány hypothesized that the drop in temperature transmitted by ice water in the ear canal caused the level of lymph fluid inside the ear to fall ever so slightly, which decreased nerve firing on that side. To the brain, this mimicked what happens when the head is turned in the opposite direction. He thought he had sorted out how the ear interacts with the brain to maintain balance.

The Bárány test became a way to see if a person had normal ear function. If the expected movements did not occur in response to warm and cold water, then it was a sign of inner ear abnormalities. Affected patients always had problems with balance.

Bárány saw that changes in lymph levels caused by temperature extremes mimicked changes in head position. He designed a chair that kept the head tilted forward as the patient was rotated to the right and to the left. He demonstrated the same eye movement pattern as seen with water infusions. He proposed that these eye movements were due to changes in inner-ear lymph-fluid levels but caused by different head positions, which in turn influenced nerve firing used by the body to keep its balance. The difficulty with this hypothesis was that it could not be proved unless the subject were placed in a gravity-free environment. Nevertheless, it seemed to work in practice, and it was for this work that Bárány was awarded the Nobel Prize nine years later. These methods of testing for inner ear function are still in use today.

Upon the start of World War I in 1914, Bárány was assigned as a medical officer to the Austrian Army at the Russian front on the border in Poland. He turned his surgical talents to treating bullet wounds to the head. The prevailing treatment had been to leave the wound open: scalp, shattered skull, and brain tissue were wrapped in gauze or simply left exposed to the air. Head wounds were left open because if the scalp were sewn over, all of the patients got brain infections and died. Nearly everyone got brain infections anyway or kept bleeding, and most of them died.

What Bárány did differently was to thoroughly cleanse the wound: he meticulously removed all skull shards, dirt, and foreign particles at the bullet entrance wound and bullet exit wound, then carefully found all bleeding vessels and tied them off. He reasoned that within hours of the bullet wound, the brain was not yet infected. He made sure to remove anything that could introduce infection and only then stitched together the scalp using a carefully aseptic technique. Out of thirteen patients, four died on the day of their operation, and the remaining nine survived. This was a significant advance in battlefield medicine.

As for trying out these unproven procedures on unwitting experimental subjects, Bárány said, "I am convinced that people with such wounds will be quite ready to cooperate in a safe and painless experiment in the interests of humanity as a whole."[1] This tradition continues today with critically injured patients sometimes receiving experimental treatments for their conditions without any possibility of informed consent. The justification is that they have a high likelihood of death in any case.

Nine surviving out of thirteen was an impressively successful statistic for that era. Bárány published his results in a German medical journal, but given that nearly the whole world was against the Germans in the war, the articles were not well read or embraced. Nevertheless, one English surgeon and one French surgeon independently published their results when they each tried Bárány's methods with some success.

In April 1915, the Russians surrounded the Austrian position at the Polish border. Bárány was taken by cattle car to a prisoner-of-war camp in Russian-occupied central Asia. The location is in present-day Turkmenistan on the northern border of Afghanistan; it really must have seemed like the end of the world to an Austrian. There was a rumor that Bárány fell ill with a malaria infection there.

1. Robert Bárány, Nobel Lecture, September 11, 1916, "Some New Methods for Functional Testing of the Vestibular Apparatus and the Cerebellum" https://www.nobelprize.org/prizes/medicine/1914/barany/lecture/.

While in prison, Bárány received a telegram informing him that he had won the Nobel Prize. This gave him recognition as a notable physician, and he was allowed to treat fellow prisoners as well as his Russian captors. He also treated the region's mayor and his family, and this might have led them to look a little more favorably on their prisoner. Meanwhile, Sweden's King Carl Gustav reached out to the president of the Russian Academy of Sciences, and with Red Cross negotiators, they worked for months to secure Bárány's release. The Russian general in charge wished to comply with the request and found a technical "out" for Bárány: the general determined that Bárány's limp was a "war-inflicted invalidism" and therefore he qualified for release. In fact, he'd had the limp ever since childhood when TB infected his bones. Finally, in 1916, Bárány was allowed to travel to Stockholm to belatedly receive his prize.

The United States joined the war in 1917, and it wasn't until American war surgeon Harvey Cushing published his results on Bárány's surgical procedures that they became widely accepted as the new way to treat head wounds. Cushing's paper did not give credit to Bárány or to any other surgeon who had explored this method. Cushing was hailed as "the father of modern brain surgery" and is still known as such today.

In 1917, when Bárány returned to Vienna, he experienced a cold reception. Despite his Nobel Prize–winning status, he did not get a professorship position at the University of Vienna as he had hoped. Instead, he was accused of plagiarism for supposedly not giving due credit to his collaborators in the discoveries of inner ear mechanisms and not acknowledging the work of previous researchers who contributed to his discoveries. The charges were investigated by the Nobel Committee and determined to be false; simultaneously, Bárány cleared his name in court. He returned to Sweden where he was welcomed on the staff at the prestigious Uppsala University.

It is necessary to keep in mind that by 1917, the United States, Japan, and Italy had joined the Allies, and the Germans were losing some significant gains they had made. Jews were enlisted men on both sides, and Jewish soldiers were as patriotic to their country of residence as any other citizens. Today, some historians estimate that twelve thousand Jews died in the service of the German and Austrian armies in World War I, and 80 percent of enlisted Jews served on the front lines. The reversal in German success called for a scapegoat, and the Jews were easy targets. Jews were accused of shirking their responsibility to fight for the fatherland or obtaining positions that were not on the dangerous front lines. The most popular accusations were that Jewish soldiers were spying for the Allies or backstabbing soldiers on their own side.

Bárány was an ethnic Jew but reported to be "nonpracticing." Nonetheless, it is likely that anti-Semitism had something to do with his reception upon returning to Vienna. Shortly after the war, Jewish names were rubbed out of war memorials. By World War II, Jewish veterans having a record of World War I service in the German Army did not garner any protection from murder because of their "race."

Bárány lived a quiet life in Sweden. His continued work on the mechanisms of balance eventually earned him a full professorship there. He suffered a series of strokes and died just before his sixtieth birthday.

The experiments that could not be done in 1906 to prove Bárány's hypothesis were finally accomplished in 1983 onboard the space shuttle *Columbia*. Warm air and cool air were alternately placed in the ears of an astronaut. When the experiment was done in the first few days of flight, it caused the same eye movements as on Earth. This was not expected. In the weightlessness of space, temperature does not conduct through fluids; the effects of hot and cold would not be what they were on Earth. This implies that something other than conduction of heat or cold on lymph fluid is responsible for the balance mechanism and eye movements. To complicate matters, when the experiment was repeated later in the mission, after the astronaut was acclimated to space, his eyes did not rotate as expected based on earthbound experiments. Leading researchers on Earth said, "We're back to the drawing board on this."

Since then, little has been written on the Bárány effect, yet the warm water/cold water test remains an entirely valid way to tell if there is inner ear disease. An article in the 2004 *Annals of the New York Academy of Science* concluded simply that things are different in space.

13

Fighting Infection

The 1919 Nobel Prize was awarded to Jules Bordet for his discoveries relating to immunity but it was not announced until 1920, after World War I had ended. Jules Jean Baptiste Vincent Bordet was born in Belgium in 1870, entered the university at the age of sixteen, and was a doctor at twenty-two. He studied at the Pasteur Institute under Ilya Mechnikov and was a world-famous medical researcher by the time he was thirty-one.

Remember from chapter 8 that Mechnikov had discovered phagocytosis, the action of the body's scavenger cells to engulf and digest invading bacteria. Mechnikov had "immunity wars" with Paul Ehrlich, who had scoffed at the idea of phagocytosis because he had already made the discovery that antibodies were responsible for immunity. It turned out that they were both right.

Eventually, everyone agreed it made sense that the body would have more than one mechanism for such an important thing as fighting infection. Paul Ehrlich had found certain biochemicals in the serum of recovered patients that seemed to have a role in overcoming infection by killing bacteria. He hypothesized that antibodies caused the production of these specialized biochemicals and thought they were customized to attack specific bacteria. He named them *complement* because they complemented the action of antibodies. Just as antibodies were designed like a lock in a key to be custom-made for the specific bacteria, Ehrlich hypothesized that the complement proteins were specific to the invading bacteria. This hypothesis of complement seemed to fit all the known facts, but it had not been proved.

Bordet and his research colleague Octave Gengou designed a set of careful experiments to sort out the nature of complement. What they found were not bacteria-specific biochemicals at all. Instead, they discovered that complement was always present in circulation and was nonspecific, just waiting to go into action when there was an infection. When complement was activated, it could be detected in serum tests.

Ehrlich picked a fresh fight over complement. The animosity he harbored was revealed in his notes to his laboratory assistants, which included these instructions: "Continue with all energy the anti-Metchnikoff [Mechnikov] matter" and "The main thing is to finish the anti-Bordet work." It was as if Ehrlich were seeking to discredit the men rather than disprove a theory or learn anything new.

Ehrlich insisted that each specific antibody had a corresponding specific complement. Bordet and Gengou believed there was only one type of complement: a sort of all-purpose infection-fighting protein. It was finally discovered that both camps were right to some extent. We now know that over twenty proteins serve as complements. Some of them work closely with specific antibodies, and some are generic free agents. Complement can be activated in the presence of an infection but is not biochemically specific to a particular strain of bacteria in the way that antibodies are.

Bordet and Gengou developed a laboratory test that would take a sample of the patient's blood and mix it with some complement taken from a guinea pig. If bacteria were present in the patient's blood, the test tube would show a pink band where the complement was interacting with the bacteria. This was a rapid and easy test to perform and revolutionized the ability to accurately diagnose infections. The test was successfully used for decades.

It was this work on immunity that earned Bordet his Nobel Prize, but it is remarkable that Octave Gengou was not considered for sharing the prize. He was coresearcher on the most significant of Bordet's work on immunity. He was five years younger than Bordet and also a Belgian. Like Bordet, Gengou was somewhat of a whiz kid, having achieved his MD degree at age twenty-two and working at the prestigious Pasteur Institute. Bordet was in the United States when notified of his prize, so he did not make the trip to Stockholm or give the traditional Nobel speech. Instead of a Nobel lecture, a staff member of the Karolinska Institute delivered a presentation on Bordet's accomplishments. Gengou was mentioned only once.

Following his work at the Paris Pasteur Institute, Bordet was named director of an infectious disease research facility in Belgium, and Gengou continued to work with him there. Eventually, it was renamed the Pasteur Institute of Brussels. In 1906, Bordet and Gengou grew the bacterium that causes whooping cough. *Bordetella pertussis* was named after Bordet. The two collaborated with others to develop a vaccine for it.

Pertussis (whooping cough) is a disease that tends to affect children. It is caused by a bacterium that releases a toxin that damages the tiny sweeping hairs that coat the lining of the lungs. This results in swelling and pain in the throat and air passages with frequent coughing spells to expel mucus. The coughing can occur ten to twenty times a day, and it can last for weeks. (The Chinese call it "hundred-day cough.") The treatment is antibiotics if it is diagnosed early. If symptoms have been present for three weeks or longer, antibiotics are not given because although the cough may persist, the body has cleared the bacteria by its own immune mechanisms. Pertussis used to happen in epidemics every two to five years. When it ran through a school, a quarter to up to a half of the children could get whooping cough. The infection rate was 75 percent to 100 percent in households of an already-infected person.

Pertussis can become a system-wide infection and even cause neurological problems with a rate of nervous system involvement from under 2 percent up to 7 percent. Building on the work of Bordet and Gengou, the vaccine introduced to prevent pertussis was made with intact, whole pertussis bacteria cells as well as with pertussis toxin. It was heavily marketed and widely injected into children even though the vaccine itself sometimes caused pertussis or severe neurological reactions, including brain damage, coma, and death.

At best, the whole-cell pertussis vaccine was only 45 percent to 48 percent effective and was only that good if the patient showed up for the entire series of five doses. Not until the late 1990s did parental outrage and scientific objections force a change in the vaccine. Today children get an acellular pertussis that does not contain a whole pertussis particle. The newer vaccine has a lower rate of adverse effects.

Like other vaccines, the pertussis vaccine is marketed to parents, public officials, and doctors on the theory of *herd immunity*. If enough people are vaccinated (the herd), then all are protected from a pertussis outbreak, even the ones who are not vaccinated. That would be a convincing argument for a common disease with no available treatment and a high rate of serious outcomes.

In fact, the recorded statistics tell a different story for pertussis. In 1838, the death rate from pertussis ranged between forty and sixty per population of a hundred thousand. Death rates from pertussis stayed about the same until 1880. For reasons unknown, the death rate began a remarkable decline at that time, which continued for decades. Just about the time that Bordet and Gengou were growing the pertussis bacterium, the pertussis

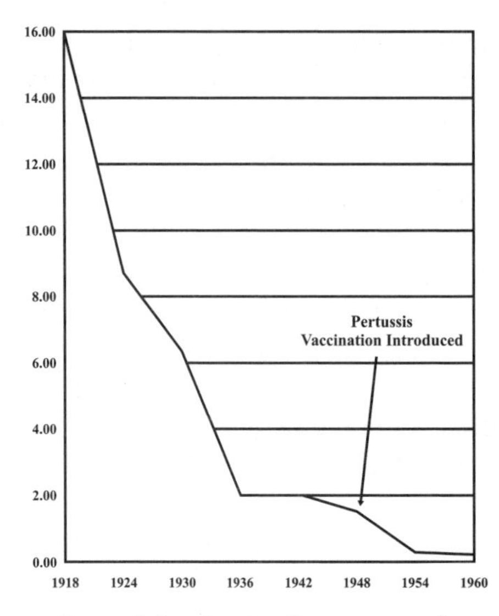

Annual death rates from pertussis mortality rates 1918–1960

death rate was continuing its steep decline; 1906 death rates were down to about twenty-five per hundred thousand. In 1917, with the introduction of the first crude vaccine, there was a spike of increased death rates for one year, but pertussis resumed its downward trend. Before pertussis vaccine was introduced for wide use in America in the 1950s, the death rate was under five per hundred thousand. This tells us that pertussis was naturally becoming a less virulent, less deadly disease on its own and completely independent of any vaccines. There have been outbreaks of whooping cough in the United States in the last decade in pockets of one hundred to two hundred children or young adults affected in various states. Ironically, the vast majority of children who fell ill had been fully vaccinated, but the vaccine effectiveness had worn off.

After 1920, Bordet became interested in yet another mechanism of immunity. Others had discovered that bacteria often produce a substance capable of destroying the very bacteria that produce it. In this way, virulent bacteria carry their own seeds of destruction. This unknown entity came to be called *bacteriophage* (literally, "bacteria-eater"). Bordet thought the mysterious agent was an enzyme that caused a chemical breakdown of the cell walls of bacteria. Others thought a virus living inside the bacteria used the bacteria to replicate itself and then burst out with the resultant destruction of its host.

Bordet was very Ehrlich-like in his scientific attacks on the alternate viral theory of bacteriophages. He also used his considerable authority as a Nobel winner and head of the Pasteur Institute of Brussels to belittle the viral theory. In 1924, American author Sinclair Lewis wrote a novel about a humble small-town doctor, *Arrowsmith*. Dr. Arrowsmith pursued his research dreams by seeking to use bacteriophages in treating a deadly plague. Lewis's character was applying the still-unproven idea that bacteriophages were viruses within bacteria. Alas, it did not work, and Dr. Arrowsmith lost his beloved wife to the plague.

The debate about bacteriophages was not settled until after World War II. The electron microscope was developed in Germany in the 1930s, and the first pictures of a bacteriophage were taken in 1939. Bacteriophages were definitely shown to be viruses living inside bacteria. The photomicrographs were published in 1940 in a German medical journal, but the outbreak of war hostilities prevented them from being widely seen. Although the Russian medical community embraced the new information, it was five years before scientists in other Allied countries understood that bacteriophages were, in fact, viruses within bacteria.

Sinclair Lewis's protagonist, Dr. Arrowsmith, sought to cultivate the virus-containing bacteria, extract the bacteriophages, and inject the virus into infected people. In fact, this was exactly what was done after 1940 in real life, in Russia and the Eastern European region of Georgia. So-called phage therapy was used extensively there, especially in military hospitals, with some remarkable successes. International pharmaceutical companies marketed phage as a treatment. In the United States, phage therapies to treat staphylococcal infections were distributed by Stanford University and the Public Health Laboratory of Michigan. Incidentally, Lewis was awarded the Nobel Prize in Literature in 1930.

Meanwhile, penicillin was invented and became widely available by 1944. Unlike bacteriophages that would kill only the specific strain of bacteria that incubated them, penicillin was *broad spectrum*, meaning it could kill many different kinds of bacteria. Antibiotics were also cheap to manufacture.

Another factor at work in the Western reluctance to use phage therapy was that shortly after World War II, the Soviet Union and Western powers fell into a mutually suspicious cold war. It pervaded all aspects of society, including medicine. All things Soviet were characterized as being tinged with Communism, including the therapeutic use of bacteriophages. As a result, the development of phage therapy was not pursued in the West.

Penicillin's success only put the idea of phage therapy on the shelf temporarily, though. Almost immediately after its release, penicillin-resistant bacteria began showing up in research laboratories. This led to the search for stronger and stronger antibiotics. By the 1990s, the occurrence of antibiotic resistance was widespread. The World Health Organization (WHO) investigation into antibiotic resistance in 2014 reported it to be a global health threat. It became obvious that an alternative to antibiotics was desperately needed, and research into phage therapy was suddenly looking more interesting. In the United States, the Federal Drug Administration (FDA) has not approved phage therapy for humans, but it is allowed for poultry and agriculture. In recent years, researchers at Rockefeller University, Texas A&M University, and biotech companies worldwide have been developing bacteriophage therapy for humans.

The various threads of this story—antibodies, complement, pertussis vaccines, and bacteriophages—serve to remind us that even the brightest minds in the history of medicine can impede research progress when they allow politics and self-importance to influence their scientific judgment.

14

Blood, Sweat, and Sugar

The 1920 Nobel Prize was awarded to August Krogh for his demonstration of how the body regulates circulation during exercise. Krogh was born in Denmark as Schack August Steenberg Krogh in 1874. In 1905, he married Marie Jorgenson, who was also a doctor and often his coresearcher.

Unlike most other Nobel winners up until his time, Krogh focused on normal physiology rather than on disease. He was interested in sorting out how the healthy body worked and sought to discover how to make it perform even better. His work focused on the small vessels (*capillaries*) that bring blood into the muscles. Capillaries form a huge network of tiny, thin-walled blood vessels that connect arteries and veins. They had only been discovered about fifty years before Krogh's prize, when they were first seen under the microscope by the Italian researcher Marcello Malpighi. Little was known about their function, for the same reason they were hard to detect. Krogh observed that when a muscle is dissected, many of its capillaries are not visible simply because they are temporarily collapsed. He demonstrated that many capillaries are collapsed when the muscle is at rest and then they open up during exercise. This mechanism allows for greater delivery of oxygen precisely when it is needed.

Another of Krogh's discoveries showed that breathing works by diffusion of oxygen through the lungs and is exchanged for carbon dioxide that diffuses in the opposite direction, to be exhaled. He also proved that fat gets burned during exercise. His discoveries led to the world's first exercise physiology laboratory and laid the groundwork for the modern-day specialties of sports medicine and physical rehabilitation.

Krogh's wife, Marie, developed diabetes within a year of her husband's Nobel Prize. Diabetes was not common in Europe at that time and there was no treatment for it. August Krogh is at least as well remembered for his role in promoting a treatment for diabetes as for his work in physiology.

Medical historians say that diabetes was first described in about 1500 BC in Egyptian scrolls and Indian medical writings. Western research had not progressed much beyond learning that the pancreas had something to do with it.

Krogh's 1920 prize gave him worldwide recognition, and he was invited to give a seminar at Yale during the couple's visit to the United States. On that visit, the Kroghs found out the pancreatic ingredient that controls blood sugar had just been discovered in a lab at the University of Toronto. *Insulin* from the pancreas pulls sugar into the cells. Diabetics do not have enough insulin, or the insulin they have doesn't work too well.

Marie Krogh desperately needed insulin, so the couple traveled to the Toronto laboratory of John Macleod and met with his insulin researcher, Fred Banting. The Kroghs obtained a license for the Toronto method of extracting insulin from animal pancreases and returned to Copenhagen to begin production immediately. They collaborated with another physician as well as a pharmacist and a drug manufacturer to cofound what is today the Novo Nordisk Company. It currently has about thirty-nine thousand employees in seventy-five countries and markets its diabetic medicines in more than one hundred eighty countries.

Thanks to insulin therapy, Marie lived to age sixty-nine. August died at the age of seventy-four in 1949.

15

The Gentile and the Jew

There was no prize awarded in medicine in 1921. The 1922 prize was shared by Archibald Vivian Hill and Otto Fritz Meyerhof for their work on muscle physiology.

A. V. Hill was born in 1886 in Bristol, England, and educated at the University of Cambridge. He developed a strong aptitude in mathematics, a subject that he enthusiastically tackled as a series of challenging games, and also gained an excellent understanding of practical physics. He turned these two disciplines toward the mysteries of muscle operation, which caught his interest because he was an avid track runner.

From his earliest studies of mathematical problems to his lifelong research into body systems, Hill's approach was that of pure fun. When challenged once about the usefulness of his research, Hill told the interviewer, "To tell you the truth we don't do it because it is useful but because it's amusing."[1] The newspaper headline the next day read, "Scientist Does It Because It's Amusing!"

Some of the things Hill did for fun were measuring precisely the nerve stimulation of muscle contraction, sorting out the mechanisms of muscle efficiency, and tracing exactly how and when oxygen was carried to and taken up by muscle cells. He won the Nobel Prize in 1922 for this work and play.

Hill was once asked, "Why investigate athletics? Why not study the processes of industry or of disease?" He explained that athletes were much easier to study because they exerted themselves for sustained periods and they could endure rigorous experiments without danger of injury. Besides, the study of athletes was fun for the researchers and fun for the research

1. A. V. Hill, *Chemical Wave Transmission in Nerve* (Cambridge: Cambridge University Press, 2009).

subjects. Seeing that it was so much fun, he hoped it might encourage young people to get into the study of physiology.

Hill was totally free with his imagination in science, sometimes proposing theories that were soon proved wrong. He embraced this outcome, pointing out that wrong theories tended to provoke interesting challenges that led to more creative research. He said, "Often I have told my young friends that when they have found something they cannot understand at all, instead of being cast down they should jump in the air for joy; for that is how discoveries are made."[2]

A. V. Hill shared the 1922 prize with Otto Fritz Meyerhof, a German born in 1884 to wealthy Jewish parents. Meyerhof seriously studied poetry, philosophy, and psychology before settling on a career in cell physiology. The experiments done by Meyerhof built on the foundation of Hill's basic discoveries and further contributed to detailed knowledge of muscle contraction. Fame from the Nobel award brought him many academic offers, including a professorship in the United States. Instead, he accepted an invitation in 1923 from the Kaiser Wilhelm Society to join its impressive research team. In 1929, Meyerhof assumed leadership of the Kaiser Wilhelm Institute for Medical Research.

Over the next decade, growing anti-Semitism permeated every facet of German society. When Hitler came to power in 1933, even people who were not card-carrying members of the National Socialist Party felt empowered to implement racist policies. For example, one of Meyerhof's colleagues at the Kaiser Institute was Richard Kuhn, head of the chemistry department. In 1933, Kuhn fired all of his subordinates with Jewish heritage. By this time, the existence of concentration camps was a known fact, and many Jews in scientific and academic positions began to flee Germany.

Meyerhof held on until 1938, when he relocated to France just a few months before the November 1938 Kristallnacht, in which Nazis rampaged in Germany by torching over a thousand synagogues and vandalizing many thousands of Jewish homes, schools, and businesses while authorities passively looked on. It is estimated that nearly a hundred Jews were killed that night and thirty thousand were hauled off to Nazi concentration camps.

In that same year, Richard Kuhn was awarded the Nobel Prize in Chemistry. Throughout the war, he continued his extensive research activities at the Kaiser Wilhelm Institute for Medical Research with a focus on

2. A. V. Hill, Trails and Trials in Physiology (London: Edward Arnold, 1965).

toxic nerve gases. He was coinventor of a poison gas called *soman*, which the United Nations classified as a weapon of mass destruction.

When the Germans occupied France in 1940, Meyerhof fled. After dodging Nazi authorities in France for a few months, he and his family were assisted by the American-based Emergency Rescue Committee (ERC). With the help of the ERC, they escaped to Spain, trekking over mountain passes of the Pyrenees and then finally making it to Portugal to board a ship to the United States. Meyerhof joined his fellow Nobel laureate A. V. Hill at the University of Pennsylvania and became a professor there.

A. V. Hill said he worked in the field of *biophysics*, the application of mathematics and physics to biological systems. The word *biophysics* was actually coined by Karl Pearson, another English mathematician. In particular, Pearson applied mathematics to the field of eugenics with the idea of breeding humans into a refined, superior race. Pearson was born Carl with a C but changed it to Karl in the 1880s when he went to study in Germany. Pearson soon became a Socialist and a noted academic Germanist (a specialist in all things German). Pearson said that any activities that helped peoples "of inferior stock" were wasted efforts and advocated that the way to improve humankind was to wage war with "inferior races." Back in England, he founded the medical journal *Annals of Eugenics* and was made the first chairman of the department of eugenics at the University of London. Pearson's racist ideas were embraced as scientific support for the policies of many warmongering world leaders, including America's Teddy Roosevelt (see chapter 6).

A. V. Hill followed in Karl Pearson's biophysics tradition but not his ideologies. Hill was a vocal opponent of Germany's racist and political policies. In 1933, Hill became the president of the newly formed Academic Assistance Council (AAC), a charitable organization that provided relocation and academic jobs for persecuted researchers who were being systematically fired in Germany because of their race, religion, or political ideas. In fact, sixteen scientists helped by the organization in the 1930s and 1940s eventually became Nobel Prize winners in various fields. After the war, the AAC's charitable activities extended to scientists and scholars in other countries suffering various forms of persecution, including refugees from Maoist China and Stalinist Russia. The organization still exists today, renamed Council for At-Risk Academics. It has also been active in Chile, the Middle East, and Africa.

Hill and Pearson attacked each other in professional publications. Pearson advocated the sterilization of mental patients and the maintenance of a registry of all of their relatives so they could be warned against conceiving children. Hill responded with an article showing how Pearson's population mathematics was defective. Pearson shot back that Hill had insufficient knowledge of mental illness to understand the problem.

In 1933, the same year Hitler became chancellor of Germany, Hill gave a lecture to international scientists at a meeting in Birmingham, England, in which he decried the concentration camps, where the German National Socialist government held a hundred thousand political prisoners. He said:

> Germany, however, has lately rendered such intellectual co-operation impossible by offending the first and most fundamental rule, that providing freedom of thought and research. . . .
>
> It seemed impossible in a great and highly civilized country, that reasons of race, creed, or opinion . . . could lead to the drastic elimination of a large number of the most eminent scientists and scholars, many of them men of the highest standing, good citizens, good human beings. This, nevertheless, has happened: the rest of the world of learning is gasping and wondering what to do about it. Freedom itself is again at stake.[3]

His comments were printed in abbreviated form in the scientific journal *Nature*. This drew a sharp response from Johannes Stark, a Nazi and 1919 Nobel Prize winner in physics. In his letter to the editor, Stark claimed there were fewer than ten thousand political prisoners, and besides, they were uniformly guilty of high treason.

A. V. Hill was a strong patriot. In World War I, he served as a captain and brevet major in the British military and was the director of the Anti-Aircraft Experimental Section. In the years leading up to World War II, he worked on a mission for the British Embassy that involved making trips to America to lure US scientists into the British war effort. (The United States did not officially join the Allies until after Pearl Harbor was bombed in December 1941.) Hill was a member of a top-secret research unit, Tizard, and his mission was to pave the way for American cooperation in refining the technology of radar, which had been developed at the University of Birmingham.

We take radar for granted these days, but in 1940, the English were being bombed regularly by Luftwaffe fleets that had escaped early detection by the

3. A. V. Hill, "International Status and Obligations of Science," *Nature* 132, no. 3347 (1933): 952–54.

radio towers dotting the English coast. For lack of radar, Royal Air Force (RAF) bombers tended to remain grounded on days when the European coast was shrouded in fog, obscuring German troop movements and the locations of munitions factories. By the time the Tizard team arrived in America with its radar in a locked black box, A. V. Hill had already had conversations with scientists at leading universities and commercial laboratories. The Tizard team found out Americans had developed their own version of radar that included devices to make it more accurate but lacked transmitter tubes of sufficient power. The exchange of technical information was perfect: England's transmitter tubes were a thousand times more powerful, and the Americans provided the apparatus that improved accuracy of English radar.

Hill's most significant contributions may have been the legacy of his kindness and his professional influences on the many researchers who were under his tutelage throughout his long professional life. For example, Dr. T. P. Feng, a Chinese scientist who worked in Hill's laboratory for three years, related a story of how Hill refused to be credited in Feng's first-ever scientific publication. Hill wanted Feng to get all the credit.

A. V. Hill gave a speech in 1952 on the need for an ethical society composed of good citizens to be responsible for the rational and constructive use of the discoveries of science. It provoked harsh condemnation from all sides. According to Hill's colleague and biographer Bernard Katz, "he noticed that denunciation followed 'alike in Pravda and the Vatican Press, so it might be concluded it was about right!'"[4] (Katz would eventually win the 1970 Nobel Prize for his investigations into the mechanism of transmitter release at the neuromuscular junction.) Hill's humanitarian viewpoints were detailed in a unique combination of logic and passion in his 1960 book, *The Ethical Dilemma of Science and Other Writings.*

---

4. Bernard Katz, "Archibald Vivian Hill. 26 September 1886-3–June 1977," *Biographical Memoirs of Fellows of the Royal Society*, vol. 24 (November 1978): 71–149.

16

The Canadian Diabetes Discoveries

The 1923 prize was awarded jointly to John James Rickard Macleod and Frederick Grant Banting for the discovery of insulin.

Descriptions of diabetes can be found in ancient medical writings of India and Egypt dating back to at least 1500 BC. Indian doctors of that era had discovered that the urine of diabetics attracted sugar ants, and called it "honey-urine." Traditional Indian medicine today is called *ayurveda*, which includes partially effective natural treatments to control blood sugar in diabetics. (Ayurveda is based on ancient Hindu Vedic texts known as the *Atharvaveda* that relate various "magical" remedies. The Vedic knowledge is traditionally interpreted as having been provided by a higher power.) Western research into diabetes did not progress for centuries. It had long been known that the pancreas, an organ behind the stomach, had something to do with digestion. In 1889, researchers at the University of Strasbourg removed a dog's pancreas to see what happened. Several days later, the laboratory's animal keeper noticed that flies were swarming around the dog's urine; sure enough, it was loaded with sugar. Thus, they knew the pancreas must contain something that controls sugar. Other researchers extracted pancreatic juice and injected it into diabetic animals, showing that something in it controlled blood sugar.

Such was the state of diabetes knowledge in the Western world in 1920 when researchers at a laboratory of the University of Toronto in Canada discovered that the substance secreted by the pancreas was insulin.

Macleod, the director of the Toronto laboratory, and Banting, the lead insulin researcher, were not getting along and were especially disagreeing about scientific methods. Their differences may have been because Macleod was a physiologist while Banting was a surgeon. Banting then allied with a relatively inexperienced student, Charles Best, to assist in his research in Macleod's laboratory. Unfortunately, the first use of their version of insulin in a human patient did not go too well: a diabetic youngster nearly died from an immediate allergic reaction.

Insulin remained in an unusable form until Macleod invited a fourth researcher into the laboratory, the biochemist James Collip. Collip took the pancreas extract created by Banting and Best and refined it into a product that could be injected into humans without creating a serious allergic reaction. Insulin is given by subcutaneous injection and drives glucose into the muscles, and in this way it lowers the level of sugar floating free in the blood.

Dr. Marie Krogh, the wife of the 1920 Nobel winner August Krogh, was among the first patients outside of the Toronto laboratory experiments to benefit from insulin. Nobel winners have the privilege of nominating people to the Nobel Prize Committee. Krogh immediately nominated Banting for insulin discovery but insisted that the Toronto laboratory chief, John Macleod, also be included. The reasoning he gave was that Banting's work relied on Macleod's earlier discoveries and on Macleod's ongoing support of his work. The committee had its doubts about whether Macleod's minor contributions were significant, but Krogh's opinion ended up convincing them.

The 1923 prize of 114,935 US dollars was awarded jointly to Banting and Macleod. Both men thought the final award was not equitable, but for different reasons. Banting ended up splitting his half with his research assistant, Best, and Macleod gave half of his share to Collip.

The isolation and purification of insulin immediately made the lab world famous and arguments arose about who should get credit. The four (Macleod, Banting, Best, and Collip) reached a truce by agreeing not to try to patent insulin independently. The patent was sold to the University of Toronto; some reports say it went for fifty cents, others say it was sold for one dollar. The idea was to make insulin a gift to humanity. However, production could not be done by an underfunded nonprofit entity, and pharmaceutical companies subsequently grew rich on the manufacture and distribution of insulin.

The insulin preparation originally isolated in the Canadian laboratory differs little from the insulin available today. Diabetic patients can be prescribed insulin extracted from a pig or cow or human pancreas or get synthetic insulin wholly manufactured by genetic engineering.

A popular alternative diabetic treatment before the discovery of insulin was a traditional ayurvedic treatment. It was an herbal combination that went by the name *diasulin*, consisting of the flower of cassia, the fruits of little gourd and bitter gourd, Indian gooseberry, turmeric root, leaves of the ram's horn plant, seeds from fenugreek and jamun, and the whole sweet broomweed plant. (This is not the unrelated natural Indonesian product trade-named Diasulin that is on the market today.)

The original diasulin was restudied in 2004 by researchers Ramalingam Saravanan and Leelavinothan Pari at Annamalai University in Chidambaram, India, who documented that diasulin controlled blood sugar in type II diabetes at least as well as a modern diabetic oral medication. Their results were very positive, so it might be puzzling that there has not been any published work on diasulin since. Of course, no one makes a profit from common plants and weeds. Instead, pharmaceutical companies are hard at work dissecting the individual active elements of traditional remedies, and worldwide some eight hundred plants have shown promise in controlling diabetes. To generate profits, the pharmaceutical industry needs to either identify a patentable extract or synthesize a look-alike from chemicals.

Insulin works by driving blood sugar into muscles. However, the human body's main use of insulin is in the liver, not the muscles. Adequate insulin in the liver regulates much more than blood sugar: it manages fuel production, cell growth, blood vessel growth, fat metabolism and storage, inflammation, storage of energy reserves, and some aspects of cancer metabolism. The available forms of insulin do not penetrate the liver. Because prescribed insulin fails to work at the level of the liver, it lowers blood sugar but does little to change the relentless course of many long-term complications of diabetes. As a result, the death rate in diabetics ages ten and older has not significantly improved in the past forty years. Meanwhile, the incidence of diabetes continues to increase.

There is some promising research by small groups not affiliated with insulin makers working on a form of insulin delivery that delivers it to the liver. Animal studies have shown good results, and if these outcomes can be translated to humans, it would be the first major advance in insulin therapy in decades.

Today there is research into diabetic medications taken by mouth that can augment insulin's effects. These need to be delivered in a form that will be absorbed. A relatively recent advance in pharmaceutical manufacturing is to make drugs in the form of nanoparticles. This concept is borrowed from the measurement known as a *nanogram*, a very tiny particle easily assimilated into the body's cells. To give you an idea of the mass of a nanogram: a paper clip weighs one gram; if divided into a thousand pieces, one piece would weigh a milligram, and if that piece were divided into a million pieces, one piece would be a nanogram.

Ancient Vedic texts on medicinal preparations called for controlled burning of the plants or plant parts and then collection of the ash to make the concoctions. We now know that this thousands-of-years-old process creates nanoparticles. When Western researchers tried to reproduce the effects of ayurvedic medicines in the past, they did not get the expected results if they failed to follow these preparatory steps. It is increasingly appreciated that the particle size of the ingredients is very important for effectiveness. Traditional ayurvedic practitioners also predicted the failure of modern pharmaceutical aims of isolating specific individual substances. They believed that combinations work holistically. Finally, Western medicine is beginning to catch on to this concept by not relying on insulin alone but giving it along with complementary oral medications that work by a different mechanism.

The world's few manufacturers of insulin have a ready market of a growing diabetic population and absolutely no financial incentive to find more effective insulin or a cure for diabetes. Instead, they have made minor tweaks in their preparations, allowing them to sequentially renew patents (a process called *evergreening*). The manufacturers are officially not allowed to engage in price-fixing, but they have all managed to steadily raise their insulin prices year by year for the last thirty years. One version of insulin that was priced at $17 a vial in 1997 sells for $138 today. Another type of insulin that has been around for more than twenty years initially sold for $21 a vial but is now $255. Sadly, the selfless humanitarian gesture by the developers of insulin therapy has been wholly invalidated by corporate profit motives. Today's diabetic patients know their brand name of insulin, but have no idea of the generous scientists who sacrificed the potential of tremendous personal profits to make insulin widely available.

17

Reading the Secrets of the Heart

Willem Einthoven won the Nobel Prize in 1924 for his invention of the *electrocardiogram*, a machine that traces the electrical conduction of the heart. Einthoven was a third-generation Dutch physician born in 1860 who contributed fundamental information to medical knowledge in diverse areas. He helped discover how the eyes perceive depth, color, and optical illusions; wrote on the mechanics of the elbow joint; studied how the airways constrict in an asthma attack; and made extensive investigations into sound perception. But Einthoven is best known for inventing an instrument to record the electrical activity of the beating heart.

Since the late 1700s, it had been established that muscles generate a small electric current when they contract. Until Einthoven's time, this knowledge was not used in patient care. In fact, the only use of electronics regarding the heart was the rare application of electric shocks to revive the apparently dead. An enthusiastic group in England calling itself the Humane Society (and later, the Royal Humane Society) worked to popularize heart-shocking treatments.

In its first annual report in 1774, the Humane Society wrote that a three-year-old girl had fallen from a first-story window and was presumed dead. Twenty minutes later, electric shocks were applied on various parts of her body, and finally, several were given on the chest. She revived in stages:

> Upon transmitting a few shocks through the thorax, he perceived a small pulsation: soon after the child began to sigh, and to breathe, though with great difficulty. In about ten minutes she vomited: a kind of stupor, occasioned by the depression of the cranium, remained for some days, but proper means being used, the child was restored to perfect health and spirits in about a week.[1]

1. William Hawes, *1774 Annual Report* (London: Royal Humane Society, 1779).

The Humane Society's goals were "to protect the industrious from the fatal consequences of unforeseen accidents; the young and inexperienced from being sacrificed to their recreations; and the unhappy victims of desponding melancholy and deliberate suicide from the miserable consequences of self-destruction."[2]

While some were running around with electric shock machines to revive the dead, others were trying to figure out the normal electrical conduction of the heart muscle. Instruments were laid directly on the surface of the exposed heart muscle of lab animals, obviously impractical for real patients. Later versions used electrodes placed on the chest, but the electrical signal appeared very faint due to the resistance and interference created by the skin and muscles.

After years of research on animals, Einthoven developed the first cardiogram that was directly useful to diagnose heart problems in patients. His original machine consisted of a thin conductive filament—the first was a silver-coated quartz thread, or string—suspended between two large electromagnets. When an electric current was detected, the string would slightly move. This movement was detected by the shadow created from a light shining on the string. These shadow changes were captured on a rolling sheet of photographic paper. Other inventors had constructed a so-called string galvanometer, but Einthoven's was many thousands of times more sensitive.

Einthoven's 1901 version of the machine was huge, weighing nearly six hundred pounds and taking up two rooms of his laboratory. The electromagnets had to be constantly water cooled. It required five people to operate. Einthoven's son, Willem Jr., was an electrical engineer who helped his father make the machine functional.

Patients in the hospital would have small metallic detectors placed on the surface of their bodies in several precise spots across the chest and on the arms and legs. These were connected to wires through which the heart's electrical signals were transmitted by telegraph cables to Einthoven's lab a mile and a half away. The telegraph signal on his end would be transmitted to the filament between the magnets and its slight movements converted to a graph tracing. The result was the characteristic regular pattern of lines and spikes we know today as the electrocardiogram (EKG).

Contemporaries described Einthoven fondly as having a kind and friendly nature. The 1922 Nobel winner, A. V. Hill, was a friend and

2. Ibid.

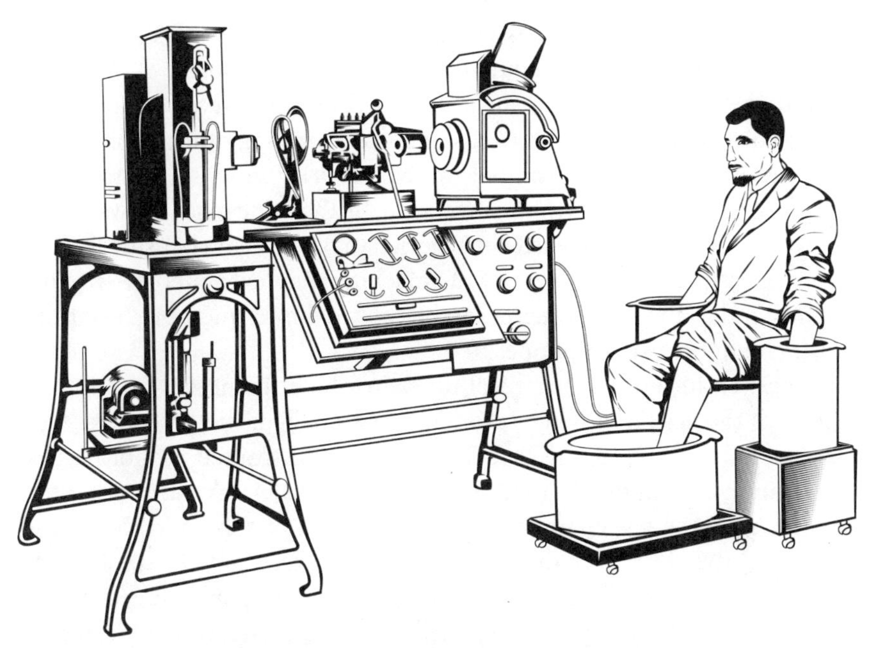

An early EKG machine built in 1911 modeled after the standards developed by Einthoven

admirer and said that Einthoven possessed grace, beauty, and simplicity. Others remarked that Einthoven was frank and direct and had great enthusiasm for his work. He rode his bike to work in the mornings and, upon arriving, would remove his jacket, tie, and stiff collar; put on his slippers; and work away until eight o'clock at night.

The only thing that upset Einthoven's normally cheerful disposition was when anyone interfered with his experiments. His laboratory at Leiden University was next door to Heike Kamerlingh Onnes's physics laboratory, which was also known as "the coldest spot on Earth" because he was trying to freeze helium down to absolute zero (minus 459.67°F). Onnes won the Nobel Prize in Physics in 1913 for his discovery of superconductivity, used in modern-day MRI machines, mobile phone bases, and particle accelerators. The next Nobel for superconductivity discoveries would not be until 1973. While Dr. Einthoven was attempting to record extremely minute electrical changes measured in thousandths of an amp, Onnes's lab next door created vibrations and electrical disturbances that interfered with Einthoven's measurements. Onnes, in turn, complained that the barking of Einthoven's laboratory dogs was disturbing him. An official university arbitration committee was required to resolve their dispute.

Einthoven left it to others to make his device progressively more practical. This involved using amplifiers and condensers, among other advances. By the time Einthoven was awarded the Nobel Prize in 1924, twenty-three years had passed since his initial invention. By then, the machine was down to fifty pounds and could easily be wheeled to the patient's bedside to diagnose a heart attack. By 1935, the EKG machine weighed a mere twenty-five pounds; today, the same tracing can be captured by a microprocessor in a wristband or on a smart phone.

To give an idea of the technical times in which the prize was awarded, 1924 was the year that King Tut's tomb was opened; the first traffic light was erected in Europe; the first fax machine transmitted a photo across the Atlantic Ocean; the railroad reached Fairbanks, Alaska; DuPont introduced rayon; the gas chamber was first used on prisoners in the United States and the electric chair replaced hanging executions in some states; the first automatic elevator was installed; the first public theaters with central air-conditioning were opened; regular transcontinental airmail service was established between New York and Los Angeles; the Boston airport opened; scientists confirmed the existence of other galaxies; dust storms were documented on Mars; and the future inventor of the CAT scan was just born.

Einthoven was on a lecture tour in America when the newspapers announced that he had been awarded the Nobel Prize. It was reported at the time that Einthoven refused congratulations, "declaring that he had had no official notification of any such honor." He was eventually officially notified and then sought out his former laboratory assistant Van de Woerd, who had long since immigrated to the United States. Einthoven intended to share his prize money with Van de Woerd in recognition of the crucial work he had done in the early years of research. Discovering his former colleague had died but was survived by two sisters who were living in poverty, Einthoven gave them half of his $40,000 prize money.

Einthoven's electrocardiogram greatly advanced the understanding of cardiovascular diseases, which were then and remain the number one cause of death worldwide. He died at the age of sixty-seven in 1927 from cancer, the number three cause of death at that time.

18

Dead Wrong

Johannes Andreas Grib Fibiger won the 1926 Nobel Prize for his theory that roundworms caused stomach cancer. Fibiger was a Danish researcher and the first of many to receive a Nobel Prize for trying to find the cause of cancer.

Cancer had been written about at least since ancient Egyptian times as evidenced by a papyrus document from 1600 BC describing a malignant tumor. The Egyptians thought cancer was caused by the gods. Around 400 BC the Greek physician Hippocrates proposed that an excess of black bile caused cancer. That theory dominated for several centuries, even though the mysterious black bile was never seen or measured. In the late nineteenth century, inflammation was raised as a possible cause of cancer, while others suggested a faulty immune system. Doctors in London noticed that chimney sweeps were more likely to get cancer of the scrotum, laying the blame on environmental toxins. In the early 1900s, doctors noticed that some of the same tumors showed up in the children of adults who had been treated for cancer, supporting the theory that chromosome damage caused heritable cancer. Breast tumors were observed to be more common in women with no children, suggesting a hormonal cause of cancer.

Fibiger noticed that laboratory rats with cancer often had parasites in their feces. He set out to discover whether the parasites caused cancer. He focused on the roundworm, which could be as long as five centimeters. He found a rich source of the roundworm parasite in the digestive tract of American cockroaches and tested his theory by feeding these cockroaches to laboratory rats. The only other food they were allowed was bread and water. Most of the cockroach-eating rats did get roundworms growing in their stomachs, about two-thirds of them had some degree of abnormal cell growth in the stomach lining, and seven of them had grown large stomach tumors. Fibiger concluded that roundworms caused stomach cancer. His discovery was widely lauded as nothing short of miraculous, and he was elected for the prize in 1926. A contemporary remarked:

> To my mind, Fibiger's work has been the greatest contribution to experimental medicine in our generation. He has built into the growing structure of proof something outstanding, something immortal.[1]

Having died of complications from colon cancer in 1928 just two years after winning, Fibiger was not around to see his work thoroughly discredited. Not all tumors are cancer, but he had declared that the tumors he saw were cancer. It turns out he was wrong; those tumors were not cancerous. It is likely the lab rats' restricted diet of cockroaches, bread, and water was lacking in vitamin A and it was merely the vitamin deficiency that promoted growth of noncancerous tumors.

The search for the cause of stomach cancer continues today. In 1984, Barry J. Marshall and J. Robin Warren discovered the bacterium *Helicobacter pylori* as a cause of gastritis and ulcers. They suspected as well that it would be found to be the cause of stomach cancer. By 1994, the International Agency for Research on Cancer declared that infection with *H. pylori*—bacteria, not worms—is a major cause of stomach cancer. This has not been proved but is based on the observation that people infected with *H. pylori* have six to eight times greater chance of getting stomach cancer than uninfected people. It does not explain how so many other people can be infected with *H. pylori* and never get cancer, or the fact that many stomach cancers are found in people without *H. pylori* infection. Marshall and Warren jointly won the 2005 Nobel Prize in Physiology or Medicine.

Even though Fibiger was mistaken, it turns out he was on the right track. Roundworms do not cause stomach cancer, but research decades later would show that a few other parasitic worms are associated with other kinds of cancer. Eating raw or undercooked shellfish in East Asia can lead to infection with certain types of flatworms (liver flukes), which are associated with an increased risk of cancer of the bile ducts. Infection with a parasite found in sewage-polluted water in the Middle East, Africa, and Asia causes bladder cancer. The so-called blood fluke can burrow into the skin and find its way to the bloodstream, then migrate to the bladder where it has sex and lays eggs in the bladder wall. The eggs can be seen in urine samples of infected patients. The theory is that the parasite infection causes tremendous inflammation, which deranges the local immune

1. W. Wernstedt, quoting A. Leitch (Nobel Prize presentation speech, December 10, 1927).

response and ultimately causes breaks in cellular DNA. The damaged DNA gives rise to cancer cells.

The Nobel Committee had considered but passed up other cancer researchers for the 1926 prize, although these were scientists whose experiments actually did show definite causes of cancer. In 1911, American researcher Francis Peyton Rous took cancerous tumors from chickens, ground them up, and extracted a liquid that had no cells in it. When this liquid was injected into healthy chickens, they grew the same kind of cancers. He concluded there was something in the liquid transmitting the cancer and speculated whether it was a cancer-causing chemical or "a minute parasitic organism" that was "ultramicroscopic." Without using the word, Rous was describing a virus, but viruses could not have been seen by even the strongest microscopes of his day. It wasn't until 1939 that the brand-new electron microscope showed an actual virus, and we now know that many viruses do cause cancer. Notes in the Nobel archives reveal that although committee members were happy to consider cockroach-eating rats, they were not impressed by work on chickens and doubted the relevance of chicken cancer to human cancer. Rous was eventually awarded the prize in 1966, more than half a century after his original research.

Another nominee who did not win was Japanese scientist Katsusaburo Yamagiwa, who showed that chemicals can cause cancer. When he painted the ears of rabbits with crude coal tar, the animals grew skin cancers. This work in 1915 became a widely used model for producing cancer in laboratory animals to study the behavior of malignancies. He was nominated several times but never won.

19

Fever Therapy and War Crimes

Julius Wagner-Jauregg was awarded the 1927 Nobel Prize for his unusual treatment of syphilis. Born in 1857, Wagner-Jauregg was an Austrian psychiatrist and major advocate of *biological psychiatry*, which was just emerging in the late nineteenth century. He was trained at the University of Vienna and later appointed a lifetime faculty position there, eventually becoming the psychiatric clinic director.

Julius Wagner-Jauregg

At the beginning of the twentieth century, Austrian psychiatrists were competing for dominance of their conflicting theories of insanity. Wagner-Jauregg's biopsychiatric approach was based on the assumption that mental distress is entirely caused by physical derangements. Sigmund Freud trained at the University of Vienna around the same time as Wagner-Jauregg and also eventually became a faculty member there. Freud took a psychoanalytic approach in which mental disorders were addressed by exploring emotional trauma, especially sexual ones. The third major player was Alfred Adler, also from the University of Vienna, who emphasized that societal stressors were the main contributor to mental afflictions.

The biological psychiatry movement got a huge boost from the discovery in 1913 that the spiral-shaped bacteria that caused syphilis could be found upon autopsy of the brains of some patients who died after experiencing mental deterioration from syphilis. This discovery was used to bolster the theory that all mental distress was from physical causes.

Syphilis is a sexually transmitted disease that starts with skin ulcers and progresses to a second stage of infection in lymph nodes and internal organs. Before the era of antibiotic treatment, up to 16 percent of untreated patients progressed to the third stage of syphilis, which affected the brain (*neurosyphilis*). It can cause confusion, depression, poor concentration, and, eventually, dementia and insanity. Old-time treatment of syphilis was mercury, but repeated dosing eventually would cause mercury poisoning, which itself could induce brain damage. In 1911, Paul Ehrlich introduced arsenic-based Salvarsan to treat syphilis, which was partially effective but also had toxicity problems. Penicillin did not arrive on the scene as a safe and effective syphilis treatment until about 1942.

Wagner-Jauregg conducted experiments on inmates of insane asylums without any informed consent as was typical of medical experimentation on mental patients of the day. His approach was based on the concept of superimposing one disease on another with the idea of somehow driving out the worse infection, or treating "evil with evil," as he put it. He tried doing this with toxin from the bacteria that caused staph skin infection and with ground-up tuberculosis bacteria, but these experiments failed. He next tried a modern version of fever therapy, or *pyrotherapy* (from *pyro*, "fire"). Wagner-Jauregg thought fever could drive out mental illness, an idea dating back to the ancient Greeks that never did work. In fact, more than three decades before Wagner-Jauregg's prizewinning experiments, Russian psychiatrist Alexander Rosenblum tried treating the insane with fever by giving them infected blood from patients who had malaria, typhoid, or relapsing fever.

Malaria brings on a pattern of recurring fevers, sometimes up to 107°F, causing the patient to become delirious or go into a coma. Victims have sudden attacks of shaking chills, severe headache, muscular aching and weakness, vomiting, coughing, diarrhea, and severe abdominal pain. Their kidneys can fail, their lungs fill up with fluid, and they can have seizures. Death occurs in up to 5 percent of malaria cases, especially if treatment is not started within the first twenty-four hours of symptoms.

In 1917, Wagner-Jauregg announced results of the study that earned him the 1927 Nobel Prize. He had injected blood from a soldier who had malaria into nine neurosyphilis patients and let them writhe with malarial fevers for up to twelve days. If they were still alive by the twelfth day, they were given quinine in an effort to treat the malaria and also given Salvarsan, the arsenic-based drug for syphilis. He reported that two promptly died, two were locked up in the asylum, and six were "improved," but four of those shortly relapsed into syphilitic dementia.

The death rate from malaria in 1917 ranged from 0.2 percent to 5 percent. In contrast, in Wagner-Jauregg's experiment, immediate fatality was 22 percent, meaning 5 to 110 times greater than the rate of death from naturally acquired malaria. Of course, that was only the immediate death rate. One ploy used to minimize the side effects of new treatments was to report only on the immediate or short-term results. Other psychiatrists began offering the treatment, and they published results that did provide longer-term follow-up data, which were disastrous. One paper provided the results of twenty-five different series of treatments totaling over two thousand patients. From 30 percent to 50 percent of syphilis patients given malaria as treatment were dead by the time three years had elapsed, with the ten-year death rate up to 78 percent.

It was already known at that time that malaria alone could cause brain damage. Before the fever occurs, the patient may suffer hallucinations, anxiety, crying, violence, and agitation. During the high fevers, the patient becomes delirious. Malaria survivors often had residual brain damage. During the height of the British Empire, physicians of the day considered malaria a leading cause of mental illness in mosquito-infested British-occupied regions such as India. A higher incidence of residual mental disorders was shown again a hundred years later in American veterans who had malaria during the Vietnam War. A very recent study of infected African children reiterated the same: malaria infection of the brain is associated with a higher risk of long-term mental derangement.

Despite the established facts, the world's highest medical honor was eventually awarded to Wagner-Jauregg anyway. It seems to have been a case precisely described by Alfred Adler when he said, "It is very obvious that we are not influenced by 'facts' but by our interpretation of the facts."[1] As with other Nobel Prizes, it is likely that politics were involved. German superiority was being heavily pushed in academic and scientific circles. According to the Nobel nomination archives released fifty years later, Wagner-Jauregg received seventeen nominations in the 1920s, but

1. Alfred Adler, *Social Interest: A Challenge to Mankind*, trans. John Linton and Richards Vaughn (London: Faber and Faber, 1938).

one Nobel Committee member in particular kept blocking the vote. Bror Gadelius, a Swedish professor of psychiatry, considered giving syphilis patients deadly malaria an outright criminal act. The election of Wagner-Jauregg only got through the committee upon the death of Gadelius. It was the first and so far only Nobel Prize in Medicine awarded to a psychiatrist.

According to contemporary accounts, Wagner-Jauregg was not a kind man. His hobby as a small child was to dissect animals. He was a card-carrying Nazi Party member, achieved on appeal; his first application was rejected because his ex-wife was Jewish. He strongly advocated for the sterilization of all persons labeled mentally ill and those with "criminal genes" and also of working women, who, he reasoned, would not be good mothers anyway. He treated masturbation by applying electrodes to shock the scrotum or with outright castration. As a military doctor during World War I, Wagner-Jauregg devised a horrifying "treatment" for shell-shocked soldiers and also for defectors from the Kaiser's army, estimated at over 150,000. He subjected them to electroshock. Historians document that field-hospital routines for treating "war neurosis" included naked exposure, burning cigarettes on patients' bodies, and applying deliberately painful electrical shocks to their nipples and genitals.

Immediately after the Great War, a law was passed in Austria authorizing prosecution for wartime psychiatric crimes. In 1920, Wagner-Jauregg was put on trial for maltreatment of patients and soldiers—specifically, abuse and torture. Among others, his Jewish colleague Sigmund Freud provided testimony that helped get Wagner-Jauregg off the hook. While Freud testified that electroshock was inhumane, he also painted a compassionate picture of a military doctor being in an impossible position in which he was forced to take extreme measures. There is no record of any coercion for the atrocities perpetrated by Wagner-Jauregg, and he did not claim any. Along with like-minded psychiatrists, he openly systematized eugenic programs well before Hitler became chancellor of Germany or führer of the Third Reich. Alfred Adler also served as an Austrian wartime doctor, and he was appalled by Freud's defense of Wagner-Jauregg's criminal brutality. It can be imagined that Freud was under some duress since Wagner-Jauregg was the head of his department at the University of Vienna.

In the early 1930s, Alfred Adler fled the persecution of Jews by coming to America, and he died in 1937. In 1938, Freud left Austria to escape the Nazis; he died in exile in the United Kingdom in 1939. Sigmund Freud was nominated thirty-three times for the prize, also blocked repeatedly by Gadelius, who called Freud's theories "illogical." Wagner-Jauregg died in 1940. None of them lived to witness the post–World War II war crimes trials held in Nuremberg. While the findings of criminality focused on the deeds committed in the course of "acts of duty" during wartime, the investigation encompassed civilian and prewar conduct as well. Therefore, it is possible that if alive, Wagner-Jauregg would have been a war crimes defendant for a second time.

The Nuremberg trials dealt with atrocities committed by the Nazi regime in the name of medical science. Criminal proceedings were held for twenty-three physicians and administrators who carried out the political objectives of the Nazis on the pretext of medical treatments, including eugenics, euthanasia, torture, and medically justified experiments. The court's proceedings heard eighty-five witnesses and evaluated some fifteen hundred documents. American judges concluded that the Nazi euthanasia program was criminal and that the often crippling or deadly experiments had obviously been conducted without consent. In response to the findings, the Nuremberg Code: Directives for Human Experimentation were composed. The directives specified that participants can consent to human experimentation only after full information is communicated and they have been given an opportunity to evaluate it; consent should be entirely voluntary and be free of coercion. The information should be imparted in a way that is comprehensible with the intention of truly enlightening the potential participant. It should include data on "all inconveniences, hazards . . . and effects on his health or person" that may come of the treatment.

The Directives for Human Experimentation grew out of unethical medical treatments administered by force and coercion, often under the guise of medical need. It was designed to protect patients and research subjects in any country no matter the political, racial, ethnic, or religious prejudices of the medical community or its leaders, jailers, judges, or doctors. It was written especially to protect persons in compromised positions such as in prisons, orphanages, and insane asylums from illegal experimentation.

Malaria therapy did not die with the introduction of penicillin to treat syphilis in 1942. It was still used well into the 1960s in studies on mostly black prisoners and patients locked up in the American South in state prisons and mental hospitals, in frank violation of the Nuremberg Code. Experiments on inmates of a mental institution in Milledgeville, Georgia, withheld malaria therapy after the experiment while the South Carolina

State Mental Hospital injected syphilis patients with a type of malaria that was known to be resistant to all drugs. In 1990, Dr. Henry Heimlich (of Heimlich maneuver fame) advocated the use of malaria to treat Lyme disease, and in 1997, he was lead author on a study conducted outside America using malaria to treat human immunodeficiency infection. This research was carried out despite the already-established fact that HIV patients infected with malaria have higher malarial death rates and malaria causes HIV to spread more readily.

As for the three competing theories of mental aberration, biological psychiatry became the most popular. Today, the psychiatric profession rides on the coattails of legitimate medicine by persistently characterizing all emotional distress and irrationality as an illness deriving from a biochemical dysfunction. The example of the syphilis bacteria infecting the brain is still used as the cornerstone of this theory. Over eight hundred conditions are fully coded for lucrative insurance reimbursement in the hefty billing bible of psychiatry called the *Diagnostic and Statistical Manual of Mental Disorders* (*DSM*). A number of biological processes such as syphilis infection or Parkinson's disease can damage the brain. Botanical or chemical substances like hallucinogenic peyote plants, amphetamines, and psychiatric drugs can cause brain damage. The resulting mental afflictions should be squarely in the physical medical realm. Instead, biopsychiatry uses these instances to continue its search for a violence gene, a bipolar virus, and any variances in the curves and contours of the brains of mental patients.

Twenty-first-century historians of the biopsychiatry movement enthusiastically laud Wagner-Jauregg as a heroic, pioneering researcher who initiated the next hundred years of biological treatments for mental conditions. His malaria experiments paved the way for the treatment of insanity with insulin coma, in which insulin is given to drive the blood sugar so low that it starves the brain and causes seizures and coma. This was followed by the development of the lobotomy and finally, deep-brain electroshock, also known as deep-brain stimulation, or DBS. Ordinary citizens are not such fans, however. In 2004, a Vienna city council member prompted an investigation into the region's civic and military heroes from the era of Nazi affiliation. This uncovered the conveniently forgotten criminal brutality, eugenics programs, and racist ideals espoused by Wagner-Jauregg and resulted in demands to remove his name from schools, road signs, and hospitals.

20

Of Lice and Men

Charles Jules Henry Nicolle won the 1928 Nobel Prize for his research on typhus. Nicolle was a French physician and researcher who studied medicine at the prestigious Pasteur Institute. He became deaf while still a young man but learned to read lips and did not let his condition limit an industrious career in researching infectious diseases. Based on his earliest personal writings, it was clear Nicolle wanted to make a name for himself in the medical research field. Luckily, the world's premier infectious disease research center was located where Nicolle had studied medicine, so he remained affiliated with Pasteur for his entire career.

To understand the circumstances of Nicolle's rise to world recognition in the field of infectious disease, it is useful to take a look at his scientific background at the Pasteur Institute. Louis Pasteur, a chemist and biologist, founded the organization in Paris and was the first to give scientific evidence for *germ theory*, which held that organisms too small to be seen were the cause of many diseases. By the turn of the twentieth century, germ theory was rapidly replacing the two other prevailing beliefs: the *miasma theory* attributed infectious diseases to some ill-defined menace in the air, and the *theory of spontaneous generation* presumed that infectious agents could arise out of inanimate objects or dead tissue without having living predecessors, such as lice growing out of dust or maggots suddenly appearing on dead flesh. Pasteur developed the first modern and effective vaccine for rabies and saved the silk industry in Europe by discovering the cause of silkworm disease. However, his most famous discoveries emerged from his research on behalf of vintners and brewers who wanted to know why their wines and beer sometimes went sour. He found that bacterial overgrowth was responsible and recommended heating the beverage to just below boiling to kill off most of the bad bacteria. The process was named *pasteurization*, and it preserved the taste of beer and wine while preventing spoilage from bacterial overgrowth. Most important to the vendors, pasteurization prolonged shelf life. France was the first country to endorse widespread pasteurization of wine, beer, and milk. Today, all mass-produced milk is subjected to ultra-pasteurization by treatment with a brief pulse of ultrahigh heat (above 275°F), which can give milk a shelf life of nine months. Unfortunately, pasteurization diminishes the content of vitamin B_{12}, vitamin C, calcium, and phosphorus.

First-generation Pastorians were colleagues and students of Louis Pasteur who fervently supported his ideas about the cause of infections, and they contributed to and furthered Pasteur's researches. Subsequently, graduates of the Pasteur Institute were often called "disciples," spreading the use of scientific methods to find germs as established by their spiritual and scientific father, Louis Pasteur. Nicolle and Pasteur barely overlapped. They met only once and Pasteur died just two years after Nicolle earned his medical degree at the institute. Thus, Nicolle was a "second-generation" Pastorian.

Pasteur disciples were already responsible for important discoveries in diphtheria, which could cause death from a sore throat in those days (brought on by toxin-secreting bacteria that resulted in a thick membrane that blocked the airway). Pasteur-affiliated researchers were already investigating cholera, which caused a sometimes-deadly diarrhea, and anthrax, mostly a disease of livestock that caused farmers to suffer severe economic losses. Other Pastorians were continuing the work Louis Pasteur had initiated on rabies, and some were researching the plague and yellow fever.

After practicing in France for a short time, Nicolle realized it would be challenging to make a prominent name for himself among the crowded field of researchers associated with the homeland Pasteur Institute, so he jumped at the opportunity to establish the Pasteur Institute in Tunisia, northern Africa. He was named its director, and there he would seek out a unique disease to study and stake his claim to fame. Pasteur outposts such as this were scattered all over the world, mostly in French colonial possessions, with the mission of bringing French culture and modern civilization to the native peoples in the form of advanced science. Today, there are Pasteur Institutes in thirty-three countries.

Nicolle arrived in Tunis in 1903 in the midst of a typhus epidemic and decided to make it the chief object of his research. Since the days of ancient Greece, typhus had been an explosively epidemic disease of unknown origin. It caused sudden severe headache, blistering fever, chills, cough, and severe pain in the muscles and joints. The skin broke out in pinpoint spots of bruising and bleeding, blood pressure dropped, and rotting gangrene could develop in the fingers and toes. The patient fell into confusion, delirium, and coma, from which the disease got its name *typhus* (from the Greek for "hazy," "smoky").

In Nicolle's day, typhus was rapidly fatal in up to four out of every ten victims. Typhus was more common in the winter months, then ebbed in the spring and disappeared in the summer. It spread like wildfire among people who lived in crowded conditions such as tent cities, asylums, prisons, military barracks, and refugee camps. Nicolle's colleague, Dr. Ernest Conseil, described the ideal breeding grounds for typhus as "filth, misery, severe weather, gatherings of men, and overcrowding." It was variously attributed to poverty, depravity, "bad air," and dogs. Contact with a typhus victim carried a high chance of infection, but it was not known if the spread was through touch, breath, respiratory droplets, or something in the air.

As a second-generation Pastorian, Nicolle hypothesized the existence of a typhus germ and started his research by observing how typhus was spread. During the 1903 Tunisian epidemic, Nicolle was supposed to accompany two colleagues to investigate an outbreak of typhus in a prison but had to cancel at the last minute. The other two doctors spent the night at the prison, caught typhus, and soon died. After the epidemic faded, Nicolle's research continued with difficulty because there were fewer human subjects. Typhus could not be easily studied in animals because animals did not naturally get typhus. In between epidemics, Nicolle needed to keep a series of animals artificially infected so he could continue to study the disease. He did this by injecting laboratory chimpanzees and monkeys with infected human blood and then transferring infected animal blood to the next animal. He did this for several years. Another Tunisian epidemic in 1909 provided fresh human cases and allowed for the observations that would eventually earn him the prize.

During epidemics, hospital workers were frequent victims, but only up to the point at which the patients underwent a hot bath before being issued a clean hospital gown. This was observed by many, including Nicolle's laboratory chief, Charles Comte, and by Ernest Conseil, the assistant hospital doctor. The conclusion was that there was something in the clothing transmitting typhus: lice. Nicolle wrote about this and claimed sole credit for the discovery, but Comte asserted that he had brought the observation to Nicolle's attention.

By the time of his Nobel speech in 1928, Nicolle made little mention of his collaborators. Instead, he spun a dramatic tale of stepping over the bodies of fallen typhus victims at the threshold of the hospital and observing that the medical staff and other patients were safe from disease once the surviving victims were bathed and given clean gowns. He said this observation was followed by a eureka moment. It was for this observation that he was awarded the Nobel Prize.

The word *lousy* comes from the description of being afflicted with lice. The discovery of the role of lice in typhus transmission led to very effective preventive measures such as making lice carriers take hot water baths, sterilizing or burning infected clothing, and cleaning hospitals, prisons, and homes with sulfur solutions to kill lice.

Nicolle initially studied lice bites on laboratory monkeys and chimpanzees. He was the first to show that guinea pigs were cheap and effective test specimens, and they went on to become the basis for experiments of many infectious diseases. Today we use the term *guinea pig* to describe anyone being tested on. Initially, Nicolle thought that the bite of the louse transmitted the infection. However, careful research by others showed it was not the bite: the louse was first infected with typhus; the infected louse excreted the germ in its stool, which looked like a gray powder; and persons with a louse bite then scratched their skin, which broke the skin and allowed lice feces to enter the human bloodstream. The lice excrement carried the cause of typhus. Nicolle confirmed this route of infection by conducting his own experiments and then took sole credit for discovering it.

Nicolle did singularly discover the phenomenon of the *inapparently infected* person. This is someone who gets infected but never shows any symptoms, yet can transmit the disease to others. The presence of inapparently infected people made the tracing of typhus sources very tricky, so it was best to undertake delousing measures in whole institutions and neighborhoods rather than focusing only on the homes of sick individuals. Some people called for systematic torching of the hut villages in rural Tunisia, which Nicolle vigorously opposed as unnecessarily savage. The 1909 epidemic was not much affected by these hygiene measures, but after it naturally wound down, the delousing measures probably served to eradicate so many lice that even the numbers of the inapparently infected went down. By 1912, there were only twenty-two typhus cases in Tunisia, six in 1913, and three imported cases in 1914.

The most famous case of inapparent infection concerned a different condition, typhoid. *Typhoid* sounds similar, but it is not related to typhus. Typhoid symptoms include fever and diarrhea. An Irish immigrant, Mary Mallon, who became known as Typhoid Mary, was a cook whose customers experienced typhoid outbreaks wherever she worked, although she never fell ill. She is thought to have infected, from 1900 to 1907, at least forty-seven people with typhoid through accidentally contaminating the food she prepared. It killed three of them. Typhoid Mary spent a third of her life in isolation as ordered by public health authorities and died in quarantine in Riverside Hospital on North Brother Island in the Bronx. An autopsy showed live typhoid bacteria in her gallbladder.

Nicolle never identified the actual typhus germ. In 1916, the typhus bacterium was finally found by Henrique da Rocha Lima, a Brazilian doctor. While researching typhus in Germany with his Czech colleague, Stanislaus von Prowazek, both doctors became infected. Lima recovered but von Prowazek died of typhus. An American researcher, Howard Taylor Ricketts, had discovered the cause of a related disease called *murine typhus*, which is transmitted by lice traveling on mice or rats. Ricketts also died of typhus. Lima named the newly discovered typhus bacterium *Rickettsia prowazekii* in honor of his two fallen comrades. Lima was not mentioned in Nicolle's Nobel lecture, nor was Lima ever recognized with even a nomination to the Nobel Committee. This was typical of the racism in the exclusively European and North American scientific cliques of the era.

Nicolle continued his typhus research by trying to create a vaccine. He noticed that most people who were lucky enough to recover from an episode of typhus were thereafter resistant to infection. It was presumed they had developed antibodies against the *R. prowazekii* bacterium. Nicolle mixed the serum of previously infected (now immune) persons with some of the bacteria and injected this mixture into himself; it was a common laboratory tradition among researchers to use themselves as experimental subjects. The idea was to provoke the immune system into making its own antibodies. Since he did not fall ill, he presumed the concoction was relatively safe and next injected children. There is no information on where he found these children, but they were likely from orphanages or mental institutions where there was no one to advocate for their rights or welfare. This was common in his era with the absence of anything like modern-day informed consent. Nicolle wrote, "You can imagine how frightened I was when they developed typhus; fortunately, they recovered."[1]

Nicolle never did find a workable vaccine. Polish researcher Rudolf Stefan Weigl later created a technique that made large-scale vaccine production possible. During World War II, Weigl's vaccine production was carried out in Nazi-occupied Poland in a laboratory that required human "feeders" the lice could bite. The bitten skin was then disinfected so the people did not get typhus. Each infected louse was dissected and the contents of its guts extracted. The lice gut was mixed with phenol and diluted in a solution, then given to humans as a vaccine to prevent infection with typhus by administering three shots of ever-increasing doses. This provoked the body to gradually develop resistance.

Most of the "feeders" were Polish intellectuals who were persecuted under the Nazi occupation; especially numerous were professors from the Jan Kochanowski University. Weigl managed to considerably assist the Jews. He covertly employed some disguised Jews and helped the Jews who were restricted by occupiers to subsist in crowded, lice-ridden ghettos. The lab personnel were left relatively alone by the SS, which wanted the vaccines for their troops but were fearful of catching typhus by getting too close to the laboratory. Weigl watered down the vaccines he prepared for the German Army but provided full-strength vaccines to smugglers who distributed it in the Jewish ghettos.

The discovery of the mode of transmission through a scratch allowing lice feces to enter the blood was not Nicolle's discovery, and he did not develop a successful way to make a vaccine. He did steadfastly continue research on other infectious diseases. He studied Malta fever, caused by bacteria growing in unpasteurized milk; contributed to knowledge on scarlet fever, tuberculosis, measles, and influenza; and studied rinderpest, or cattle plague, a deadly viral infection of hoofed animals that is related to the measles virus (rinderpest was finally eradicated from the world in 2010). He discovered the parasite that caused toxoplasmosis, a common gut infection found in the stool of some 11 percent of Americans, most of whom are without symptoms. It usually did not cause symptoms until the person's immune system was compromised by HIV infection or cancer

1. As quoted by Ludwik Gross in "How Charles Nicolle of the Pasteur Institute discovered that epidemic typhus is transmitted by lice: Reminiscences from my years at the Pasteur Institute in Paris," *Proceedings of the National Academy of Sciences*, vol. 93, 10539–10540, October 1996.

or immune-suppressant medications. At that time toxoplasmosis infection could become active and result in death.

Nicolle was competing for recognition from the Nobel Committee among many worthy researchers at the various Pasteur Institutes and beyond. He desperately wanted the esteem the Nobel Prize would confer and sought to be nominated. Directly approaching the Nobel Committee was forbidden, but Nicolle urgently lobbied his scientific colleagues to nominate him. In fact, he was nominated thirteen times in six years but was repeatedly disappointed. Finally, in 1928, he received inside word that he was highly favored for the prize, and this time his expectations were not disappointed. The prize was specifically for his typhus research done almost two decades earlier. As mentioned, he did not deign to give significant credit to anyone else in his Nobel speech. It is possible that the committee was considering this a prize for Nicolle's lifework in the field of infectious disease in general, even though the rules read that it should be given only for a specific contribution, not for a body of work. In spite of the doubts about Nicolle's claim as a sole discoverer of some things, his total output on the topic of various infectious diseases was large and varied, even by today's standards.

Beyond his medical research, Nicolle was very dissatisfied by how things were run at the Pasteur Institute in Paris and the influence it held in its outposts around the globe. He lobbied the father Pasteur Institute to break its long-standing tradition of never selling human pharmaceuticals at a profit. Nicolle advocated for the organization to make financial arrangements with pharmaceutical manufacturers to produce the vaccines it developed, such as for anthrax and rabies. His personal papers revealed that he sought financial gain primarily for himself and his collaborators and secondarily for the Pasteur Institute. He independently entered into contractual agreements with the French drugmaker Poulenc to make vaccines and antitoxins from raw substances he provided. Pasteur Institute researchers would subsequently develop vaccines for tuberculosis, yellow fever, diphtheria, tetanus, polio, and hepatitis B. Today, the Pasteur organization remains nonprofit but does contract with pharmaceutical giants to produce vaccines; all monies support its research institutions. Nevertheless, mismanagement has brought the Pasteur Institutes nearly to bankruptcy more than once in the last century.

Nicolle also violated tradition by divorcing his wife and establishing a liaison with the widow of his deceased laboratory partner, who had contracted a mysterious illness in Nicolle's laboratory and died. Nicolle wrote three fictional stories using flowery language to explore philosophical questions such as the happiness of man, the beauty and savagery of nature, and the meaning of desire in dreams.

Nicolle long suffered with heart problems, and on the day that the king of Sweden was to bestow the Nobel Prizes, Nicolle was too sick to attend. His speech was read in absentia. His heart troubles were exacerbated in 1934 when he fell ill with murine typhus, with which he had been experimenting in the laboratory, the kind caused by related typhus bacteria that is transmitted by mice and rats. It infected his heart and started a steady decline in his health. He died in 1936 at the age of seventy.

Today, the typhus vaccine is no longer commercially available because wide-scale use of DDT in post–World War II years diminished the lice population, routine hygiene and disinfection measures are used in most jails and military installations, and the disease is cured by taking a single dose of the antibiotic doxycycline. Typhus is always present at a low prevalence (*endemic*) in some regions of the Andes highlands and in Africa in Burundi, Ethiopia, and Rwanda. The most recent epidemic in 1997 affected between twenty-four thousand and forty-five thousand in Burundi jails and refugee camps, with a death rate of about 3.8 percent. In the last hundred years, typhus only became epidemic only during wartime, so prevention is now social and political rather than medical.

21

Hidden Vitality

The 1929 Nobel Prize was shared by Christiaan Eijkman and Frederick Hopkins, each for his work on recognizing the existence of hidden substances in food that later came to be known as vitamins. It was the first Nobel Prize in Medicine for vitamin research, although the chemistry prize the year before had recognized the work of vitamin D researchers. In subsequent years, there would be several more prizes for vitamin research.

Eijkman was a military medical officer in the 1880s, serving in Dutch-occupied Indonesia. On his first trip, he caught malaria and was sent home on sick leave but recovered enough to return to what is now Jakarta. He was assigned by a Dutch government commission to study why there was an epidemic of deadly *beriberi* in the region. Beriberi causes profound weakness, numbness, weight loss, emotional disturbance, heart failure, and death. Numbness and paralysis started in the feet and climbed up the legs. Patients began to walk with a high-stepped gait to compensate for their weak feet. Then the hands and arms became affected. The muscles wasted away as paralysis set in, breathing became difficult, and lungs and other tissues were congested with fluid. One report described beriberi as "insidious in its attack, rapid in its progress, and fatal in its termination." Progress of the disease could be rapid indeed: Eijkman described seeing a soldier perform well in target practice in the morning and drop dead of beriberi that same night.

Beriberi was first described around 1700 but became very common in Indonesia in the second half of the 1800s. The name comes from the Sinhalese word *beri*, meaning "weakness." The disease happened more often in crowded conditions such as hospitals, prisons, ships, and military camps while the free-living native population was not much prone to beriberi. In keeping with the brand-new germ theory of the era, it was presumed that beriberi was some sort of infectious disease able to spread quickly in crowded settings. The search was on for the beriberi bacteria or parasite.

Dutch military doctors before Eijkman had managed to find some bacteria in the bloodstream of soldiers who had died from it, and Eijkman used that to infect monkeys and rabbits. They did not get ill right away, so he switched to using an animal with a faster life cycle: the chicken. He injected some chickens with the organism and soon they did get the weakness typical of beriberi, but so did the chickens that were not injected with the supposedly infected blood. It was logical to presume the infected birds spread it to the others, so his next experiment separated the cages, but all of the birds got beriberi this time too. This turned the search away from looking for an infectious agent because something else was obviously going on. To make matters more confusing, the sick chickens unexpectedly recovered fully.

In his Nobel lecture forty years later, Eijkman related that his laboratory aide, responsible for care of the chickens, just happened to mention that he had been feeding them cooked, polished (white) rice from the military-base kitchen, but when a new cook took over, he "refused to allow *military* rice to be taken for *civilian* chickens." Instead, all chickens were now getting ordinary "rough" (brown) rice. The direction of research abruptly changed, and Eijkman started to look for a disease-causing toxin in the white rice.

The cultivation of rice dates back to before 2500 BC and rice has fed more people than any other crop. *Polished rice*, as it was called then, is simply rough rice that has had its outer hull-like layer rubbed and shaken off. The outer layer contains oil, which makes rough rice prone to turn rancid when stored; in contrast, polished rice can be stored for long periods without spoiling. To further rice production in 1870, the Dutch brought in a new method of processing rice: the steam-driven mill. This made the production of polished rice faster and cheaper. Milling also reduced the bulk of raw rice by a third, making polished rice more economical to store and ship. Suddenly, white rice was dominant.

Other researchers working in Asia had already discovered that an exclusive rice diet was the culprit in beriberi. Eighty-five years before Eijkman's chicken experiments, the first detailed English description of beriberi was made in Ceylon (present-day Sri Lanka) by a British army physician, Thomas Christie. Christie highly suspected that beriberi was due to lack of a nourishing diet, implying that a white rice diet was lacking an essential nutrient. In 1832, another British military surgeon, this one serving in

India, likewise concluded that beriberi was due to a nutritional deficiency caused by a rice diet. In 1879, just before Eijkman arrived in Indonesia, a Dutch navy doctor, Frederik Johannes van Leent, observed that native sailors were less likely to get beriberi if they ate a more varied European diet. Van Leent concluded that an exclusive rice diet was creating a nutritional deficiency but could not identify the missing factor. Van Leent's missing-factor idea was rejected by the Dutch commission, which held to the prevailing view that sufficient calories of carbohydrates, fat, and protein provided all that was necessary for health. They did not have our modern-day knowledge of the necessity of additional nutrients in minute amounts (*micronutrients*).

Eijkman also may not have known about the work of Dr. Kanehiro Takaki, who had tackled the same problem a few years before on Japanese naval ships. Takaki found that enlisted men who ate mostly white rice were more prone to beriberi than officers and he suspected a deficient diet. All he could tell was that the common enlisted men often ate rice exclusively, while officers enjoyed more variety in their meals. By 1880, Takaki correctly concluded that a heavy rice diet resulted in nutritional deficiency and narrowed the "missing something" down to a substance containing the element nitrogen. He showed that the risk of beriberi in sailors could be eliminated by varying the all-rice diet to include nitrogen-containing foods, especially proteins. Takaki was in fact zeroing in on the missing micronutrient, which was indeed made partly of nitrogen. He convinced the navy officials to institute dietary changes for the common enlisted men, virtually eliminating the beriberi problem in the Japanese Navy.

Meanwhile, in 1885, Eijkman was toiling away at still trying to figure out what he suspected was a toxin in polished rice. His next experiment fed polished rice to one group of chickens and rough rice to the other. As expected, all of the polished rice eaters started falling over and dying, but the birds fed on rough rice stayed well. From his subsequent experiments, he concluded there was something in white rice that caused beriberi, possibly a toxin, and something in the brown rice hulls that treated beriberi. He called it "the antineuritic factor." Unlike Christie, Takaki, and van Leent, Eijkman did not grasp the concept of a vital nutrient at all. Instead, he thought brown rice provided some sort of remedy to the excess carbohydrates in an exclusively white rice diet.

Eijkman was recalled to the Netherlands, and his replacement in Indonesia was Gerrit Grijns. Grijns looked over Eijkman's data and correctly concluded that white rice was not toxic, but that it lacked some vital ingredient contained in the hulls of rough rice, the same conclusion Christie, Takaki, and van Leent had made years earlier. To check whether the chicken results would be the same in humans, Grijns enlisted the help of a prison medical inspector on the islands. A survey of nearly three hundred thousand inmates in 101 prisons revealed that the rate of beriberi was three hundred times greater when the staple diet was polished rice. Recommendations to change prison diets were ignored, but it was not only the prison population that was affected. Schools, orphanages, hospitals, mining camps, plantation lines, and military bases all relied on a white-rice diet and suffered high rates of beriberi. Nevertheless, these facts went largely unrecognized.

Some twelve years later, in 1898, the world-famous Scottish physician Patrick Manson wrote about beriberi patients in British India:

> They lie like logs in their beds, unable to move a limb or perhaps even a finger. Some are atrophied [wasted away] to skeletons; others are swollen out with dropsy [puffy from water retention]; and some show just sufficient dropsy to conceal the atrophy the muscles have undergone.[1]

Manson, who was called "the father of tropical medicine," regarded beriberi as just another strange part of the unsolved "tropical pathological puzzle" and concluded that it was still one of many "diseases of undetermined nature."

Over the next twenty years there were many similar reports about the relationship of beriberi to polished rice from virtually all rice-eating nations that had been studied: China, the Philippines, British India, Siam (Thailand), and French Indochina (Vietnam) were heavily affected by beriberi. In each of these locations, the European occupiers had likewise replaced traditional rice preparation methods with automated milling, which gave them the advantage of being able to store and cheaply ship white rice.

In 1910, Japanese researcher Umetaro Suzuki isolated thiamine from rice hulls and proved it was the elusive antineuritic factor sought by Eijkman twenty-five years earlier and was in fact the nitrogen-containing vital nutrient proposed by Takaki even earlier. Suzuki called it *aberic acid* (from *a*, "against," + *beric*, "weakness"). If the diet is lacking in thiamine, then the nerves do not function well and they rapidly degenerate. He reported his

---

1. David Arnold, "British India and the 'Beriberi' Problem, 1798–1942," *Medical History* 54, no. 3 (July 2010): 295–314.

findings in a Japanese medical journal, but when it appeared in a German journal, the translator had not understood the part about thiamine (aberic acid) being the newly discovered hidden nutrient, and this fact was omitted from the translated paper. That same year, a recently formed professional organization, the Far Eastern Association of Tropical Medicine, brought its official dietary recommendations to the heads of governments: polished rice as a staple diet should be supplemented; alternately, they recommended, it should be switched out for rough rice.

Not all nations heeded the advice, and the delay in taking action cost millions of lives. In 1913, a researcher from the Rockefeller Foundation's International Health Board estimated there were one hundred thousand deaths per year from beriberi in Asia. As late as 1930, a report on beriberi by a public health commissioner in India made no mention of vitamins but focused on the possibility of a toxin in the soil of rice fields.

In 1911, Polish biochemist Casimir Funk also looked for an antineuritic factor in rice hulls, not knowing about the discovery of thiamine by Suzuki the year before. Funk isolated a different substance, niacin, and coined the new word, *vitamine*, to describe it (from *vit*, "life," + *amine*, "a compound containing nitrogen"). By definition, vitamins are substances needed from outside the body that the body cannot manufacture internally. It turned out that although niacin was an important nutrient in other ways, it did not prevent beriberi. Ironically, we now know that niacin is not even an actual vitamin because the body manufactures its own niacin from the amino acid tryptophan. The general public still considers niacin a vitamin, and it continues to be marketed that way along with other B vitamins.

The full structure of the nutrient Suzuki had found in 1910, thiamine, was not determined until 1935, seven years after Eijkman's Nobel Prize and fifty years after his initial discoveries about polished versus rough rice.

Eijkman was nominated several times for the prize starting in 1914 and finally won it in 1929. He was too ill to attend the award ceremonies and died the following year at the age of seventy-two. Not only did he not heed the work of earlier researchers who correctly deduced that beriberi was due to a vital nutritional deficiency, he never actually discovered a vitamin. Five years in advance of Eijkman's chicken experiments, careful observations by Takaki led him to the fact that beriberi was a nutritional deficiency caused by a vital missing factor that was composed of nitrogen, and it was Suzuki who found the actual vitamin. The oversights in the 1880s were understandable, but by the time of the prize in 1929, all of the facts were known. Suzuki was nominated by his Japanese colleagues for the Nobel Prize in 1936 but did not win; a Japanese national would not win a Nobel Prize in medicine until 1987. Today, thiamine is known as vitamin B1, the first B vitamin to be discovered, and it is routinely added to white rice to prevent beriberi.

The cowinner in 1929 was Frederick Hopkins, an English physician with primary training in chemistry. He was a pioneer in the new field of *biochemistry*—the study of how chemistry is a part of biological life. Hopkins was a contemporary of Eijkman and likewise received the prize for research he had carried out decades earlier. In his Nobel lecture, Hopkins reflected on the view of nutrition in the late 1800s when it was thought to be just a matter of energy in/energy out: a certain amount of calories supplied energy for all bodily functions. With this simplistic formula, the field of nutrition was considered finalized with nothing more worth investigating. But even then, there was ample evidence that the simple formula did not hold up in real-life experiments on living organisms, animal and human alike. For example, an experiment in 1881 fed one group of mice milk, while another group of mice was fed each of the components of milk (the carbohydrates, the protein, and the fat). The component-fed mice died while the mice fed on regular milk survived. This led to the conclusion that although adequate calories were essential for life, they were not all that was required to survive. The work for which Hopkins received his share of the Nobel was conducted a quarter of a century later in 1906 to 1907. He simply repeated the experiments on milk-feeding components, different in that he used rats instead of mice, rats being closer to humans in their physiology. Hopkins's experiments were rigorous, using a large control group with meticulous measurements of rate of growth for the survivors and rate of decline for the dying group.

It became evident that some compounds were required by the body in such small amounts that they did not contribute to energy supply. In other words, these micronutrients were necessary for health no matter how many calories were eaten. By 1911, using Casimir Funk's invented word, all agreed to call these substances *vitamins*. (Many sources erroneously cited Funk as the discoverer of vitamins.)

By the time of his Nobel lecture in 1929, Hopkins said, "The complexity of these nutritional needs as we now view them is indeed astonishing." Although he and Eijkman were getting the world's highest honor in medicine and the very first Nobel Prize in the field of vitamin research, neither of them actually discovered a vitamin. Hopkins humbly admitted their contributions were among the work of many others. In his Nobel lecture, Hopkins said, "Who was the 'discoverer' of vitamins? This question has no clear answer."[2] Hopkins went on to make it clear that Casimir Funk was not the discoverer of vitamins and diverged to counter the public comments made by Funk that had thoroughly minimized Hopkins's contribution to vitamin research. This was a rare use of the Nobel lecture to publicly strike at a colleague.

The problem in giving this first prize for vitamin research was an artifact of the Nobel rules. They required that nominations be made only to individuals, and a shared award could go to three at most (or an organization that the nominated individuals represented). The many different vitamin researchers were from a variety of institutions around the globe, too many to recognize in a single Nobel Prize.

Hopkins made many other tremendously important discoveries, including isolating the amino acid tryptophan. Tryptophan is a building block of the neurotransmitter serotonin and of niacin—the fact that niacin is made by the body from tryptophan-containing foods is what disqualifies niacin as a vitamin. Like a vitamin, tryptophan is needed for survival, but humans cannot make it. It is called an *essential amino acid*. Hopkins also discovered glutathione, today known to be the body's most active antioxidant. Glutathione prevents cells from suffering damage from the by-products of metabolism and from harmful substances we consume.

2. Sir Frederick Hopkins, "The Earlier History of Vitamin Research (Nobel Lecture, December 11, 1929)," Nobel Lectures in Physiology or Medicine 1922–1941 (Amsterdam: Elsevier, 1965).

22

Of the Type That Saved Millions of Lives

Karl Landsteiner won the 1930 Nobel Prize for discovering the major blood groups A, B, and O, making safe transfusion possible. Landsteiner was another Nobel winner brought up in the University of Vienna system (like Bárány and Wagner-Jauregg), where he was first a student and then a faculty member in the late 1800s.

Although blood transfusions had been tried since the 1700s, they caused serious reactions and oftentimes death. Many transfusion deaths followed a complex reaction involving fever and chills, muscle pain, nausea, pain in the chest, belly, and back, blood in the urine, and yellow jaundice.

A 1900 publication by Karl Landsteiner contained an important footnote: "The serum of healthy human beings not only agglutinates animal red cells, but also often those of human origin, from other individuals." [1]Landsteiner observed that when donor blood was mixed with patient blood in a test tube, sometimes the patient's blood cells stuck together and clumped up and sometimes they did not. This process was called *agglutination*. In 1901, he used himself and five doctors in his lab to do an experiment, finding that when a person's blood was mixed with a drop of his own blood, no reaction occurred. Blood samples from two of the doctors did not react with each other but agglutinated when mixed with a drop of blood from the third doctor or the fourth doctor. Blood from two other subjects, including Landsteiner, did not cause agglutination to anyone's blood, but blood from anyone else would cause the blood of these two to agglutinate. Landsteiner drew up a chart of these confusing results and realized he was dealing with the immune system rejecting blood that was

1. Karl Landsteiner "Zur Kenntnis der antifermentativen, lytischen und agglutinierenden Wirkungen des Blutserums und der Lymphe," [Knowledge of the antifermentative, lytic and agglutinating effects of blood serum and lymph], *Zentralblatt für Bakteriologie, Parasitenkunde und Infektionskrankheiten* 1900; 27:357–62.

of a different "type." He concluded that each person had a specific type of red blood cell and assigned one type as "A," another as "B," and the last one that didn't provoke agglutination with any other patient's blood as "O." In addition, he concluded that those with blood type A had something in the liquid part of the blood—the serum—that agglutinates type B cells, and those with type B blood agglutinate type A cells.

He thought these agglutinating factors were a kind of antibody, which was a word that had been coined by Nobel winner Paul Ehrlich in 1891. This meant that people with type A blood could get transfused with blood from another type A, but they had antibodies that would destroy type B blood cells. Likewise, people with type B blood could get cells from another type B but would have an antibody reaction to type A cells. The serum of type O patients agglutinated type A cells and type B cells, so people who were type O could get blood only from other type Os, but people with type A and type B blood could get blood cells from type O donors. Think of the type Os as smooth, naked, red blood cells that have nothing on them for the type A or type B serum to react to. People with type O blood are *universal donors*—anyone can use their blood without suffering an antibody reaction. This explained why some patients had survived blood transfusions—they'd been lucky enough to get blood from a donor who just happened to match their blood type or were a type O donor.

In his 1901 paper describing these results, Landsteiner concluded: "My observations reveal characteristic differences between blood serum and red blood cells of various apparently healthy persons . . . the reported observations may assist in the explanation of various consequences of therapeutical blood transfusions."[2]

A colleague of Landsteiner discovered the fourth blood type the following year: people with type AB blood can receive blood from type A, type B, and type O but can give blood only to others with type AB. So people with type AB blood are *universal recipients*—they can be transfused safely with blood from any type.

Initially, Landsteiner's landmark discovery met with only lukewarm interest. Minimizing its utility, 1908 Nobel Prize winner Paul Ehrlich asked what sense it would make since "things are in the blood circulation directed against quite heterogeneous material which under normal circumstances

2. Karl Landsteiner, "Über Agglutinationserscheinungen normalen menschlichen Blutes," [On agglutinitive phenomena of normal human blood] *Wiener Klinische Wochenschrift*, 1901;14: 1132–1134.

can never come into the picture."[3] In truth, Landsteiner himself had not fully grasped the lifesaving benefits of his discovery, meekly expressing the hope that his observations might somehow help explain the various consequences of blood transfusions.

Practical application of these discoveries was not appreciated until Landsteiner and others did many more experiments proving that blood typing held true in all people. It was not until about 1909 that blood matching was regularly used in hospitals, but it turned out that incompatible blood types were not the only reason blood clotted and caused fever and chills in the patient. Even matched blood would often clot, sometimes in the collection chamber or tubing. This problem was partially solved by rigging up a tube connecting the donor's artery to the patient's vein directly, which exposed the blood only minimally to an artificial environment. But the arrangement required considerable surgical skill and was not practical for everyday transfusions, and many patients died from blood clotting anyway. Glass collectors and glass tubing were tried, which did not work any better than rubber tubing. In those days, the transfusion apparatus was reused from patient to patient, and it was finally observed that if the apparatus was cleaned very well using triple distilled water, then chills and fever were about ten times less likely. It was not until 1914 that a tiny amount of *sodium citrate* (a simple chemical that prevents blood clotting) was added to collected blood. Sodium citrate is still used today to prevent banked blood from turning into one huge clot. Understanding the need to match donor to patient blood type along with the use of citrate to prevent clotting finally made blood transfusions easy, fast, and relatively safe.

Landsteiner and others continued research into blood types, finding that the A, B, and O designations were only *main* blood types; there were many more *minor* blood types that could be detected for more exact transfusion matches. Years later, Landsteiner was the codiscoverer (with Alexander Wiener) of the reason a newborn baby sometimes developed a destructive anemia shortly after delivery. The anemia was due to another antibody related to red blood cells called the Rh factor.

Like those of so many of the early Nobel winners, Karl Landsteiner's research interests were diverse. He identified the proteins that were activated to cause a life-threatening allergic reaction. He showed that the spirochete bacteria that caused syphilis could be more clearly seen with a

3. F. Himmelweit, M. Marquardt, H. Dale (eds), *Collected Papers of Paul Ehrlich*, vol. 2 (London: Pergamon, 1957), 313.

dark-field microscope that brilliantly showed light bacteria against a black background. He took part in experiments that demonstrated polio was transmitted by something in the patient's spinal cord, later shown to be a virus.

Landsteiner was born to Jewish parents just outside Vienna. His father died when Karl was only six. At the age of twenty-one, Landsteiner and his mother converted to Catholicism. He was raised by his mother and stayed with her as a single man through her death in 1908, not marrying until eight years later. His biographical entry on the Nobel Prize website describes that for the rest of his life, Landsteiner kept a death mask of his mother on his wall. The death mask was an eerie fad in nineteenth-century Europe, made by taking a cast of the deceased's face, thus preserving the lifeless features of the dead as a cherished relic. Death masks date back to at least ancient Egyptian times, and such masks were made of Peter the Great, Beethoven, Stalin, and Nikola Tesla.

There had been a Jewish presence in Austria since at least the twelfth century. An official order to annihilate Jews had been given by Albert V in the early fifteenth century. By the nineteenth century, there was less resistance to Jews. In the year before Karl Landsteiner was born, the Austrian constitution of 1867 was modified to allow Jews to live and practice their religion in Austria. In the next forty-five years, warring among Austria's border neighbors and then World War I increasingly drove Jewish refugees into the country; this strain fueled a reemergence of anti-Semitic sentiment. By 1916, Communist prerevolution was brewing in Russia, and the Austrian Christian Socialist Party propaganda of the day identified Austrian Jews with the so-called Bolshevik threat. In 1918, immediately after World War I, the Christian Socialists urged the German-Austrian people "to take decisive defensive action against the Jewish menace," paralleling overt anti-Semitism in neighboring Germany. In 1919, when Austria reorganized into the Republic of German-Austria, many people of means with Jewish heritage decided to flee. Landsteiner went to the Netherlands and then to America where he secured a research position at the Rockefeller Institute.

Even from a position of relative safety in America, Landsteiner is reported to have lived in fear of a Nazi takeover and persecution for his own Jewish heritage. In the 1930s, he filed for an injunction to prevent a biographical listing in *Who's Who in American Jewry*, explaining in his petition:

> It will be detrimental to me . . . to emphasize publicly the religion of my ancestors; first, as a matter of convenience; secondly, I want nothing that

> may in the slightest degree cause any mental anguish, pain or suffering to any members of my family. . . . My son is now 19 years of age and he has no suspicion that any of his ancestors were Jewish. I know as a positive fact that if my son were to see the book that is about to be published it would be a shock to him and might subject him to humiliation.[4]

Landsteiner's fear was not unfounded considering the political climate. His employer, the Rockefeller Institute, was well documented to have funded German eugenics programs. In 1929, Rockefeller funded building of the Kaiser Wilhelm Institute's Institute for Brain Research and continued to support it well into the 1930s while it was run by Ernst Rüdin, the leader of the psychiatric designers of Hitler's racial genocide program. Rüdin advocated "private and government force" to remove undesirable people from society. Hitler himself awarded Rüdin the Eagle Shield of the German Reich in recognition of the pseudoscientific rationale he provided to support the Nazis' systematic murder of Jews, gypsies, psychiatric patients, anyone labeled crazy, alcoholics, the mentally challenged, people with epilepsy or certain other neurological conditions, prisoners of war, or those who held dissident political views. The most notorious funding from Rockefeller went to the laboratory of Otmar Freiherr von Verschuer, where Josef Mengele worked before he became the infamous Auschwitz death camp doctor. A prominent scientist working at the Rockefeller Institute at the same time as Landsteiner was Alexis Carrel, the 1912 Nobel winner and a vocal eugenicist. In his 1935 book *Man, the Unknown*, Carrel had advocated the efficient disposal of unwanted persons by means of gas chambers, and within a few years they were in use by the German Reich.

The day after *Anschluss* (the official annexation of Austria into Germany) in March 1938, Jews in Vienna began to experience open, systemic brutality. Despite its significant ongoing support of eugenics programs inside Germany, the Rockefeller Foundation also funded rescue programs for politically persecuted scientists, mostly Jews, first in the 1930s with the Special Research Aid Fund for Deposed Scholars and then with the Emergency Program for European Scholars until 1945. In the 1940s, most of the remaining Jewish population in Vienna became victims of the Holocaust. Landsteiner died in 1943 in the safety of America.

If Albert Einstein is the acclaimed peerless genius among the physical scientists of this century, who can claim the equivalent title among the biomedical scientists? To answer that question, the research contribu-

4. Howard M. Sacher, *The Course of Modern Jewish History* (New York: Random House, 1958)

tions of twenty-four Nobelists in medicine (contemporaries of Einstein) who received the Nobel Prize between 1912 and 1966 were analyzed by a medical historian. When assessed on the criteria of influence in multiple disciplines, revolutionary opening of new vistas of knowledge, and significant impact on human life, Karl Landsteiner's discoveries of human blood groups were judged to be comparable in stature to Einstein's discoveries.

The impact of Landsteiner's discoveries also influenced a number of biomedical disciplines such as immunochemistry, medical anthropology, forensic medicine, genetics, and pathology.

23

Greatest Influence from the Lowest Profile

Otto Warburg won the Nobel Prize in 1931 for discovering the exact biochemical reactions by which cells take up oxygen and use it to make energy. Warburg was a German research biochemist and physician. He came from the well-known Warburg family whose members first made their mark as bankers in fourteenth-century Venice. When Venetian rulers restricted banking activity by Jews, the family operations moved to Bologna and then to Germany, where they founded the city of Warburg and continued banking. Extended family members branched off into the sciences, arts, and medicine, as well as politics. Subsequent generations of the German Warburgs gradually dropped traditional Jewish names such as Moses and Abraham in favor of classic German names such as Siegmund, Felix, and Otto. Otto Warburg's father converted to Protestantism and married a non-Jewish German, as was quite common in nineteenth-century Germany.

Otto Warburg never practiced medicine by seeing patients but instead strictly kept to laboratory research. The only exception was his volunteer service in the German Army in World War I, in which he served in a medical unit. He was awarded the Iron Cross when he was injured in the line of duty and then allowed to return to his laboratory. Unlike many scientists of his day, he also shunned taking an academic position or working where he would be required to spend extensive time teaching, which would have drawn him away from his beloved laboratory. He never married, on the advice of his mother, who warned that it would distract from his scientific pursuits. His mother arranged for retaining a capable young man to act as Otto's personal assistant, with whom he lived until the end of his life.

Warburg's initial interest was solving how cells got energy, seeking to know what made them live and grow. To study normally invisible chemical reactions at the cellular level, he had to develop revolutionary new biochemistry research methods, many of which are still in use today. His early research identified the remarkable similarities between chlorophyll in plants and hemoglobin in animals (including humans). The chlorophyll molecule is a complex ring structure with magnesium in its center, while hemoglobin has nearly the same ring structure but with iron at its center. Chlorophyll is used by plants to take up carbon dioxide and release oxygen into the air, while hemoglobin is used by animals to carry oxygen and ultimately release carbon dioxide.

Warburg's Nobel Prize was for discoveries he had made many years earlier in identifying how cells take up oxygen and use it to make energy, generating leftover carbon dioxide at the end of the reaction. This can be thought of as how cells breathe and is technically called *cellular respiration*, as it parallels how the lungs work. Subsequently, Warburg's main interest was sorting out how the metabolism of cancer cells was different from that of normal cells. Again, this work required Warburg to invent brand-new techniques for preparing cells and measuring their ultrafast, minute chemical reactions.

Warburg was the first to demonstrate that while a normal cell took up oxygen and used it for energy, the cancer cell preferred to take up sugar for energy, generating lactic acid as a waste product of the sugar metabolism. He noticed that this acidic environment somehow empowered cancer cells to grow at a mad rate. Most remarkably, cancer cells took up sugar to metabolize even when oxygen was around. When oxygen was lower than normal, uptake of sugar increased, and when oxygen was cut off, the use of sugar for energy went into high gear. This has since come to be known as the Warburg Effect. Warburg concluded that cancer cells had something inherently wrong with their regulation of respiration since they preferred sugar even when oxygen was available. He hypothesized that a defect in cancer-cell respiration was the primary cause of cancer. His main findings regarding cancer-cell respiration were published in 1931, the same year that he received his Nobel award.

Warburg announced these findings in an era of intense scientific competition to discover causes of cancer. In fact, he was nominated for the Nobel forty-nine times by the time he won and had seriously been considered for the Nobel in 1926 when it went instead to Johannes Fibiger for the assertion that stomach cancers were caused by parasite infection, later shown to be false. Warburg's cancer theories were too controversial for the Nobel, so it was awarded for his discovery of the mechanism of normal cellular respiration.

For the rest of his life, Warburg studied the defects in cancer cells that made them take up sugar instead of oxygen. He researched the cancer-promoting effects of radiation, food additives, tobacco, other pollutants, and the effects of nutritional deficiencies. In a brief movie clip recorded at the 1931 Nobel awards, Warburg was seen to be smoking, although he soon turned into an antitobacco advocate and kept a personal organic garden to ensure the purity of the foods he consumed. He helped to prove many of the cancer-generating effects of various toxic substances but thought that emphasis on such research was detracting from his main work: finding out how a normal cell's respiration became corrupted into using sugar metabolism instead of oxygen metabolism.

His exclusive biology laboratory at the Kaiser Wilhelm Institute was initially wholly financed by the German government. The year of his prize was also the year the Rockefeller Foundation funded two new laboratories at the Kaiser Wilhelm Institute, and Warburg was assigned as director of the new Cell Biology Lab there. In April 1933, two months after Hitler was elected by the National Socialist (Nazi) Party, a new civil service law prohibited Jews from working in government-funded positions. All Jewish scientists were fired from the Kaiser Wilhelm Institute except Warburg and a few who held director positions. By this time, Warburg was widely recognized as a world-class, illustrious scientist with 207 published articles and two books.

By 1937, other departments at the institute were actively "researching" racial purity and methods of mass murder, apparently fully known by the institute's board and funded in part by the Rockefeller Institute. A distant relative, Max Warburg, was kicked off the board and fled from the country, as did most of the extended Warburg family. But Warburg remained untouched and made no effort to leave. He was dismissed for three weeks in 1941 but soon reinstated by intervention of Hitler's Reich Chancellery. Hermann Göring, Hitler's cabinet minister, who established the Gestapo, "reassessed" Warburg's ancestry as only one-quarter Jewish, which qualified him to continue to be employed. Some historians feel this extraordinary intervention was due to Hitler's known obsessive fear of cancer.

Warburg continued to doggedly work in the lab, staying in relative safety and maintaining high productivity. From 1934 through the end of the war, he published another 105 scientific articles. His work was not interrupted again until the Russians marched in and forced him to temporarily relocate outside Berlin. After the war, he was severely criticized for continuing to work under Hitler. However, no war crime charges were brought against him as he never did human experimentation or engaged in any unethical research activities. It appears that the German Reich did not profit from any of his work. At least three Warburg extended family members who remained in Germany died in concentration camps.

Internationally, scientists also cold-shouldered Warburg for a time after the war, but interest in his outstanding research soon overcame their reservations. His laboratory in Germany was visited often by Nobel winners and other stellar researchers whose own work complemented and furthered Warburg's theories. He was so highly regarded that he was nominated two more times for the Nobel for later discoveries—in 1944 (in medicine) and 1960 (in chemistry). He went on to publish three more books and another 191 scientific papers.

Warburg was totally willing to defend scientific principles even when they were politically unpopular. For example, in the mid-1960s, Warburg spoke up for Dr. Josef Issels, a German cancer doctor who is regarded as the "father of integrative medicine" for his holistic therapies. Issels was heavily attacked by German conventional medicine. He was also convicted of manslaughter by a high court because some incurable cancer patients died under his unconventional care. Warburg offered to testify at Issels's appeal, but the convictions were overturned and his testimony was not needed.

Warburg was seriously criticized for his insistence that the primary cause of cancer was a defect in cellular respiration. In his final publication in 1970, the year of his death, he opined that the switch to cancer-cell respiration was due to either a lack of enough oxygen or a lack of vitamin B_1 (thiamine) and advocated vitamin therapy for cancer treatment. Despite the criticisms, cancer research has continued to rely on many of the discoveries made by Warburg. One reliable index of the significance of individuals' contribution to worldwide scientific advance is the number of times their work is cited in medical journal articles. Warburg's papers have been increasingly cited in modern research publications at an almost exponential rate in the four decades since his death.

24

The Brain Is Not the Mind

The 1932 Nobel Prize was awarded jointly to Sir Charles Scott Sherrington and Edgar Douglas Adrian for their discoveries regarding the functions of nerves.

Sherrington was born in London in 1857 and worked mostly in England, but for a short time, he studied cholera outbreaks in Germany, where he worked under Robert Koch, the 1905 Nobel winner.

Sherrington conducted experiments in the late 1890s that radically changed the understanding of how nerves function. Until then, it was thought that each nerve fired in an isolated *reflex arc*, with a stimulus causing a firing of a single nerve that in turn caused a muscle to twitch. Instead, Sherrington showed that for every nerve activation there was a corresponding inhibition of an opposing nerve. For example, when the nerves to contract the biceps are fired, the nerves to inhibit the triceps must also fire because the biceps can contract only when the triceps are made to relax. His discoveries were made possible by the work of many others, including the meticulous nerve drawings of Ramón y Cajal and the nerve-staining technique of Camillo Golgi, the 1906 Nobel winners.

Expanding on this concept of nerves working along multiple communication lines with each other, Sherrington published additional studies in 1906 that further changed how the brain and nerves were thought of. He showed that the peripheral nerves, spinal cord, and brain work in a complex interacting system (*the integrated nervous system*), with intricate patterns of nerve stimulation and nerve inhibition. He identified three specific types of sensory channels. The nerve endings that sense the outside environment detect light, sound, smell, and touch. Nerves lining the mouth, gut, and lungs sense the internal regulation responses, such as taste, breathing, hunger, heart rate, and the need for bowel movements. The third type consists of nerve endings that maintain an upright posture against gravity and let us know the position of our body parts. Sherrington was the first to make a highly accurate map of which nerves sensed the skin. For example, sensation in the thumb is felt by way of the nerve branching off the spine at the level of the sixth neck vertebrae, the index and long fingers are sensed through the seventh cervical nerve, and the ring and baby fingers are felt via the eighth cervical nerve. Every medical student memorizes this map. Sherrington's discoveries led to improved methods for rehabilitation after nerve injury and made brain surgery more of a sure thing, although it was still very risky in his day.

Sherrington invented a word for the junction between nerves, calling it a *synapse* (from Greek *syn*, "together," + *hapsis*, "joining"). This started many other researchers on the path of investigating how the impulse jumped the gap. A contest arose between "sparkers," who thought signals were propagated by electrical charge, and "soupers," who thought a chemical passed the signal across. A couple of soupers (Otto Loewi and Sir Henry Hallett Dale) earned the Nobel Prize in 1936 for discovering the first chemical neurotransmitters.

Sherrington's laboratory work was severely interrupted with the advent of World War I. At age fifty-seven, he was too old for soldiering, so he went to work in a munitions factory seven days a week for three months, where he gained firsthand experience of "industrial fatigue." In 1916, he advocated for women to be admitted to the medical school at Oxford. Oxford had been in existence since about the year 1100. Although women had studied at Oxford since the 1870s, they were not allowed to graduate and receive certificates. It was not until 1920 that women were allowed to graduate and the action was made effective retroactively—any woman who had taken all courses and examinations in the past was allowed to return to Oxford for an official matriculation.

Nerve and brain research inevitably led to speculations on the relationship of mind and brain. Many of Sherrington's contemporaries were looking to the physical sciences to answer the eternal questions of philosophy like "How do we actually think and know about ourselves?" They sought a specialized form of energy called "the mind" and tried to determine whether the mind was a physical thing in a part of the brain. Sherrington wrote extensively on this topic. In his 1933 Cambridge lecture "Brain and Its Metabolism," he denied a "scientific right to join mental with physiological experience." He argued that we would never find the mind in the physical brain. In his book *Man on His Nature* (1940), Sherrington detailed all of the considerable knowledge of brain anatomy and functions and described how the brain could not possibly be the mind. He wrote: "As followers of natural science we know nothing of any relation between thoughts and the brain, except as a gross correlation in time and space" and "[The mind] goes therefore in our spatial world more ghostly than a ghost . . . it is a thing not even of outline; it is not a 'thing.'"[1]

Sherrington did not find any basis for a physical brain structure or location for things such as the finer aspects of personality, detailed memory, reasoning, ethics, and faith. He was content with the unknowns of the mind, stating in *Man on His Nature*, "We have to regard the relation of mind to brain as not merely unresolved but still devoid of a basis for its very beginning."[2]

Sherrington was not bothered by the fact of the mind, a nonphysical entity, coexisting with the physical brain but not part of it. In the foreword to his 1906 book *The Integrative Action of the Nervous System*, he wrote, "That our being should consist of two fundamental elements [physical and psychical] offers I suppose no greater inherent improbability than that it should rest on one only."[3]

This was a subject of intense interest in an era when so many new facts were being revealed about the workings of the brain and nervous system. Many of the most illustrious neurophysiologists of Dr. Sherrington's day concurred with him. Sir John Eccles was once a student of Sherrington's and became a Nobel Prize winner in 1963 for his work on the nerve synapse. Eccles wrote extensively on the differences between the mind and the brain. In his book *How the Self Controls Its Brain*, Eccles described the primary controlling force as being the "I," the sense of self, which is not composed of energy but influences energy and physical things. Another colleague and world-renowned neurophysiologist, Dr. Wilder Penfield, extensively studied the workings of the brain in search of the human soul and concluded: "The brain is a computer, but it is programmed by something outside of itself."[4] Another contemporary neurophysiologist, F. M. R. Walshe, had a lighter look at the whole matter and conceded that humans

1. Charles Scott Sherrington, *Man on His Nature*, The Gifford Lectures, Edinburgh (Book 1937), 2nd ed. (Cambridge: Cambridge University Press, 1951).
2. Ibid.
3. Charles Scott Sherrington, *The Integrative Action of the Nervous System* (New Haven, CT: Yale University Press, 1906).
4. Wilder Penfield, *Mystery of the Mind: A Critical Study of Consciousness and the Human Brain* (Princeton, NJ: Princeton University Press, 1975).

are more wonderful and dignified than science can unravel. He said, "In the last resort they [doctors] are concerned with the ultimate particular thing, the human person, whom science by itself is inadequate to comprehend."[5]

Dr. Sherrington lived to the age of ninety-five. He remained bright and cheerful and was lauded with ninety honorary degrees and fellowships throughout his long, productive career.

The 1932 cowinner was Edgar Douglas Adrian, also an Englishman and neurophysiologist. Adrian took over the laboratory of his former mentor, Keith Lucas, who was killed in a training flight during World War I. Adrian continued Lucas's work to study nerve impulses. They were such tiny electrical charges that they were difficult to measure, on the order of a few microvolts lasting for a few thousandths of a second. Adrian solved this by using a minuscule meter to detect the electrical discharge of a single nerve hooked up to a cathode ray tube (such as in an old-fashioned television set), which amplified the nerve impulse. He was able to show nerve firing set off by pressure, touch, and temperature demonstrated five thousand times bigger than the original. In the same way, Adrian measured the return nerves firing back from the brain that ultimately caused muscle movement.

Adrian observed that sensory nerve cells discharged a whole series of impulses. In each fiber, the nerve impulse was of a constant size: the stronger the stimulus, the greater the initial frequency of the impulses discharged along the nerve. He showed that a stimulus of constant intensity applied to the skin immediately excited the nerves, and the frequency of nerve impulses (how many impulses per second) increased as the stimulus was increased. For example, when a finger was pinched, there might be nerve excitation impulses sent at a frequency of ten per second. This frequency was relayed to the brain as pain. When that pinch increased to a crush, the frequency of nerve impulses increased to fifty per second. The brain registered stronger pain as relayed by the increased frequency of the impulses. Adrian found that sensory impulses passing along the nerves were initially constant in strength, but as the duration of the impulses continued, they reduced in frequency, and as a result, the sensation in the brain began to weaken.

After winning his Nobel Prize, Adrian focused his research on accurate mapping of the electrical activity of the brain. His work popularized the

5. F. M. R. Walshe, "Thoughts upon the Equation of Mind with Brain," *Brain* 76, no. 1 (March 1, 1953).

concept of a *homunculus* (literally, "little man" in Latin). By mapping which brain regions perceived which sensations, Adrian developed a cartoonlike drawing of a distorted human figure perched in an awkward position over the sensory regions of the brain. The homunculus had its feet toward the middle of the brain, its shoulders on the edges over the feet, an upside-down head, then large lips, hands, feet, and genitals to correspond to the increased sensory fibers in those body parts. This detailed new information made it possible to do selective brain surgery in a relatively safe manner.

Adrian's research stopped abruptly in 1958 when a water leak damaged all his laboratory equipment. He continued to teach and lecture, eventually became the chancellor of the University of Cambridge, and then served as chancellor of the University of Leicester.

The discoveries of Sherrington and Adrian remain foundations in the understanding of the nervous system. In subsequent decades, at least thirty more scientists would get the Nobel Prize for discoveries built on the fundamentals established by these two.

25

American Genesis

Thomas Hunt Morgan won the 1933 Nobel Prize for genetic research. He was the first American to win the prize in medicine and it was the first prize for genetic research.

Morgan was born in Kentucky in 1866. When he completed his medical studies in the 1880s, Morgan became interested in the science of heredity. He studied the work of his predecessors Charles Darwin and Gregor Mendel.

In 1859, the Englishman Darwin advanced the theory of *survival of the fittest*, which supposes that the challenges in the outside environment cause the most advantageous characteristics to be inherited, while organisms with useless or hindering characteristics eventually die out.

Gregor Mendel was a German monk who grew crops of pea plants to see how plant characteristics were passed from generation to generation. His landmark paper was published in 1866, the year Morgan was born. Mendel proposed that the inherited substance determining plant characteristics could be dominant or recessive; a tall plant crossed with a short one had offspring that were always tall, tall being dominant and short being recessive. Mendel also showed that the inherited factors would remix in the next generation of plants, which yielded tall plants and short plants in a ratio of three to one.

Neither Darwin nor Mendel knew what the cellular substance was that passed on traits and caused heredity. But earlier in the century, about 1840, biologists had already discovered some material inside the nucleus of cells that picked up dye. These were named *chromosomes* (from *chromo*, "color," + *soma*, "body," literally, "dye-bodies"), but no one knew their function. Around the time Morgan was graduating from medical school, chromosomes were postulated to carry the mysterious inheritance factors that were later named *genes*.

To study inheritance patterns efficiently, Morgan abandoned plants and animals in favor of fruit flies. The *Drosophila* species (commonly called banana flies) had a fast reproductive cycle. They were initially raised on clumps of bananas in the laboratory but soon grew even faster on lab dishes containing sugary nutrients. Also, fruit flies had only four pairs of chromosomes, making them easier to study compared to a human's twenty-three pairs. Morgan and his three laboratory assistants at Columbia University raised thousands of flies and studied their traits, such as wingspan, eye color, and body length. One day a mutant white-eyed fly was hatched, which was lucky because this was such a noticeable trait that it was easy to mate this fly and follow the inheritance patterns.

Morgan's laboratory assistants were graduate students Alfred Henry Sturtevant and Calvin Blackman Bridges, and they did most of the tedious recordings and calculations. It was Sturtevant's statistical charts that revealed the gene for the white-eye trait was always inherited along with the gene that determined the sex of the offspring. This led to the realization that some traits were reliably found to be spatially near or adjacent to others and they always were inherited together. It was through an analysis of their detailed statistics of inheritance patterns that the team proposed that genes were ordered along a chromosome like beads on a necklace, evenly spaced, and that the position of certain genes could be pinpointed. These discoveries initiated the modern era of genetics research. Their theory was finally proved much later with electron microscope pictures of the chromosome.

It was more than two decades after the fruit fly hatchings that Morgan received the Nobel Prize for the groundbreaking work conducted in his laboratories. In his acceptance speech, he recognized the work of his assistants Sturtevant and Bridges, and he shared his prize money with them. A third lab associate, Hermann J. Muller, would go on to earn his own Nobel Prize in 1946 for the effect of X-rays on genetic mutations.

Like many other geneticists of his day, Morgan was at one time a member of the Eugenics Records Office (ERO). The ERO was located in Cold Spring Harbor, New York, founded in 1910 by Harvard biologist Charles Benedict Davenport and managed by Princeton graduate Harry H. Laughlin. The organization was initially supported by grants from railroad heiress Mrs. E. H. Harriman, the Rockefeller family, and Dr. John Harvey Kellogg (of breakfast cereal fame). In later years, the ERO was funded largely by the Carnegie Foundation.

The ERO used the emerging field of genetics to support racial superiority policies. They conducted door-to-door surveys and supposedly showed that the vast majority of Russians, Hungarians, and Jews living in America were "feebleminded" and used their findings to advocate for forced sterilization of persons they deemed socially inadequate for breeding. They argued for laws against mixed-race unions for fear of diluting the racial superiority of whites and supported strict limitations on immigration. Epileptics, alcoholics, the mentally challenged, and members of any non-Aryan race were all considered inferior and in need of mandatory birth control.

Morgan withdrew his support of the American eugenics movement in about 1915 and quietly urged other scientists to do the same but was not public about his feelings until writing his 1925 textbook, *Evolution and Genetics*. There he criticized the eugenics movement for being more based on propaganda than on science. Throughout the 1930s, Hitler's National Socialist Party in Europe drew on many of the ERO studies and further pseudoscientific support from papers written by American eugenicists. By 1939, the Carnegie Foundation began to feel the heat for its funding of ERO, conducted an investigation, and was "shocked" to find that the science for eugenics was so flawed. The foundation withdrew its funding, and the ERO was closed. Morgan lived until 1945, long enough to see how pseudoscience could prop up an amoral regime such as Nazism.

26

Eat Your Liver

The 1934 Nobel Prize winners in physiology or medicine were all Americans. The prize was shared three ways by George Whipple, George Minot, and William Murphy for their discoveries in the treatment of a particular type of anemia.

George Hoyt Whipple was born in New Hampshire to an old American family with ancestors documented on New World soil dating back to the 1600s. He was a relative of William Whipple, a signer of the Declaration of Independence. His father and grandfather had also been doctors, but the young Whipple was more interested in physiology research than in seeing patients. After establishing himself as a world-class scientist at Johns Hopkins in Baltimore and serving as dean of the University of California medical school, he was invited to become dean of a new medical school in Rochester, New York (later the University of Rochester). It was there he conducted the studies that would win him the prize.

Whipple systematically studied the effects of various diets on anemic laboratory animals. The results were highly variable depending on the species studied and the exact breeds. His longtime research assistant, Frieda Robscheit-Robbins, supplied the lab with English-Dalmatian bull-dogs, which provided consistency for making reliable comparisons when diets were changed. Beginning in 1925, the two published a series of papers on "Blood Regeneration in Severe Anemia," reporting on the effects of a long list of foods tested on animals. They established that a poor diet could cause anemia, and certain foods could effectively treat anemia.

Anemia means low red blood cell count. Most people are familiar with the two most common forms of anemia: anemia due to blood loss (such as bleeding from a gunshot wound or an internal tumor) is treated by blood transfusion; anemia due to lack of iron (such as can occur in preg-nant women) is treated with iron-rich foods or iron shots or pills. Whipple focused his studies on pernicious anemia. *Pernicious* is from the Latin word meaning "destructive" (from *pernicies*, "ruin," based on *nec*, "death"). It is the function of red blood cells to carry all-important oxygen into the body's tissues, and when there are fewer blood cells (anemia), there is less oxygen delivery. In pernicious anemia, the red blood cell count goes down gradually and the effects are slowly but profoundly destructive. Before Whipple's research in the mid-1920s, pernicious anemia was utterly untreatable and often fatal.

Symptoms of anemia are tiredness, shortness of breath, and dizziness, but if it comes on very gradually, these symptoms may not be present. Anemia can cause headaches, cold hands and feet, heart palpitations, an enlarged heart, and heart muscle failure. Pernicious anemia in particular is not caused by low iron but by some other missing factor in the diet, not yet discovered in the 1930s. In addition to the generic symptoms of anemia described above, this kind of anemia causes nerve damage, with numbness and tingling in the hands and feet, muscle weakness, and loss of reflexes. The patient becomes unsteady and loses balance easily. The bones also become weak, and hip fractures are likely. Severe pernicious anemia affects the brain with confusion, depression, dementia, and memory loss. The patient can be nauseated and vomit frequently, with loss of appetite and weight. A smooth, thick, red tongue lacking the normal crevices and furrows is characteristic of pernicious anemia. Infants who have pernicious anemia can be irritable and have facial tremors as well as difficulty with their tongue and mouth movements, thus difficulty with breast-feeding. If not corrected, these infants may have permanent growth problems.

Whipple found that pernicious anemia generally improved in laboratory dogs fed on meat rather than fruits or vegetables with one exception: apricots could reverse anemia faster than beef heart or beef muscle. However, the champion treatment was found to be raw beef liver; it had by far the most impact on blood count, reversing pernicious anemia in as little as ten days. The liver contains iron, but iron alone did not reverse pernicious anemia. Whipple and his assistants tried cooking the liver down to ash and feeding this iron-rich substance to the dogs, but it did not completely treat pernicious anemia. The same foods were also known to be high in B vitamins, so again they fed the anemic animals on a few of the known B vitamins, but these vitamins had only a partial effect.

Whipple and his laboratory assistants eventually conducted research for the pharmaceutical company Eli Lilly to develop a liver preparation

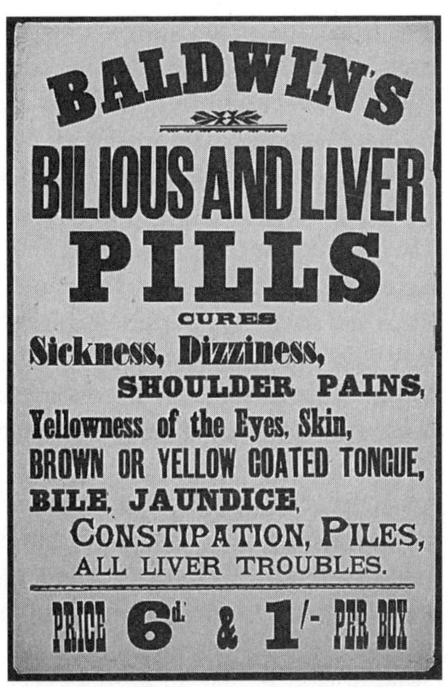

Vintage advertisement for liver pills

that could be put in an injection or pill form. They failed to capture the unknown but crucial extract and could not duplicate the results obtained from simply eating beef liver. The team of Whipple and Frieda Robscheit-Robbins published twenty-one papers on the topic, and their landmark paper identifying the superiority of liver treatment listed her as the first author. First-author position is usually given the lion's share of credit for a discovery, yet Robscheit-Robbins was not nominated for the prize. In fact, no woman had yet been nominated for the Nobel Prize in Medicine. When Whipple was awarded the prize, he gave Robscheit-Robbins much credit in his acceptance speech and shared his third of the prize money with her.

In Whipple's ninety-five-year lifetime, he made many contributions to our understanding of the liver. He discovered the liver makes *fibrinogen* (a clotting substance), and it is similarly affected by diet. Fibrinogen promotes normal clotting, but extremely high levels are associated with premature hardening of the arteries and subsequent heart attack and stroke. Diets high in iron, sugar, and caffeine will increase fibrinogen. He was also the first person to describe a mysterious condition affecting middle-aged men with diarrhea, weight loss, and neurological decline, which is today called Whipple's disease. His first case was seen in 1907 in a medical missionary. Whipple did an autopsy and correctly assumed the disease had a bacterial cause, but the responsible bacterium was not identified until research on the tissues of Whipple's other disease patients in 1992. In 2003, researchers were able to obtain the preserved tissues from the autopsy Whipple had conducted on his original 1907 patient. The tissues showed that the 1907 patient was indeed infected with the bacterium that has subsequently been found in patients with Whipple's disease.

Whipple found that the liver has an amazing capacity to regenerate itself after injury from an accident or surgery or chemical damage. Today it is known that 75 percent of the liver can be removed or damaged, and yet it has the potential to grow back to full size.

Whipple discovered that dogs deficient in protein would suffer liver injury from anesthesia drugs that are routinely administered during surgery and that the liver is protected from anesthesia damage if the animal is on a high-protein diet or simply given one of two amino acids (L-methionine or L-cystine) before anesthesia. He showed that laboratory dogs could be kept alive on an intravenous (IV) drip of essential nutrients including the amino acids, completely bypassing the processes of eating and digestion. These studies were later performed on humans and led to the development of IV nutrition for surgical patients or any patient who needs to rest his bowel. This method is used today for patients with bowel diseases, severe malnutrition, and excessive weight loss.

In the 1920s, the recent invention of the X-ray was increasingly being used in medical settings, but the upper limit of safety was not known. Whipple and his research colleagues documented the effects of a single lethal dose of radiation on dogs, remarkably similar to what happens in humans: swelling, redness, and inflammation of the digestive tract leading to vomiting, diarrhea, and then death within four days. He was well ahead of his time. Today physicians recommend keeping track of one's lifetime dose of medical radiation, particularly from CT scans. In 2011, the Institute of Medicine issued a report detailing that half of all excess radiation comes from medical X-rays, especially CT scans, leading to extra cancers of the breast, lung, and brain, among other organs. Up to 30 percent of CT scans are not medically necessary. At least some cancer risk could be avoided.

Contemporaries of Dr. Whipple only related professional descriptions of him, as he did not believe in the mixing of academics with personal details or social life. While dean at Rochester, Whipple felt that medical students should not date or be married lest it detract from their studies. He wished to be remembered as a teacher.

When the papers of George Whipple and Frieda Robscheit-Robbins were published, George Minot and William Murphy were struggling with how to treat pernicious anemia in people. They learned of Whipple's work and tried their patients on raw liver, which was not terribly palatable. Then they tried cooked beef liver, which worked well, reversing the anemia in a short time. They even recorded complete cures. This became known worldwide and suddenly pernicious anemia ceased to be a deadly condition.

Although Minot was nominated for his role in the cure for pernicious anemia, he assisted with another notable but unrelated discovery. Dr. George Minot served as a surgeon for the United States Army in World War I and held a special interest in blood. He was recalled to duty in 1918 to help discover what caused illness in workers at a munitions plant in New Jersey, where casings were filled by hand with melted TNT crystals. TNT is an explosive that is more stable than Alfred Nobel's dynamite and harder to detonate, and it makes a less-powerful explosion. Several workers developed blue hands, lips, and gums, yellow eyes, purple pinpoint bruises, weight loss, and death. If removed from the assembly line when symptoms first developed, the workers could fully recover, but if they kept working, they could be dead within a week from the first sign of sickness. Minot's colleague in the project was Dr. Alice Hamilton, who would later become known as "the mother of occupational medicine" for her specialty in discovering toxic exposures in the workplace and for making factories safer places. The illness in the munitions factory had a component of anemia and liver failure, so Minot's insight was needed in the investigation at the factory. Hamilton deduced that skin contact with TNT was the poisoning route.

Minot and Hamilton both became Harvard professors (she was Harvard's first female professor). Minot was diagnosed with diabetes in 1921 at the age of thirty-five. Even though insulin was discovered that very year, it was not yet widely available outside the research laboratory, and the only treatment for diabetes at the time was severe calorie restriction. Most diabetics died of complications within a year or two. Minot's doctor put him on a diet of 530 calories a day. He was six feet one inch tall and lost weight down to 135 pounds. In 1922, insulin became available and Minot was saved. Eventually, he suffered from diabetic complications and had a stroke, dying in 1950 at the age of sixty-four.

The third cowinner was William P. Murphy. He was interested in medicine but could not afford tuition, so he worked as a high school teacher until he could scrape together funds for the medical school at the University of Oregon in Portland. Despite taking a side job as a laboratory assistant, he ran out of money by the end of his first year and was forced to drop out. He enlisted in the army for the last two years of World War I, then found out about a scholarship to Harvard Medical School sponsored by a Harvard alumnus with the same last name. William Stanislaus Murphy wanted to fund education "for men of the name of Murphy."

Murphy distinguished himself at Harvard and remained in Boston as a research physician and hospital staff doctor at the Peter Bent Brigham Hospital. He perfected liver extract to the point where a shot in the muscle worked better than liver extract by mouth or eating whole liver.

Murphy was John F. Kennedy's attending physician during JFK's two-month hospitalization at Brigham in 1935 when he was eighteen years old. JFK often had health crises throughout his life and collected various diagnoses, including colitis, ulcers, emotional stress, severe arthritis, low thyroid, adrenal failure, low blood counts, abscesses, urinary tract infections, and prostate trouble. When young JFK's blood count dropped, Dr. Murphy gave him a series of liver injections that did not improve his condition. His low blood count must not have been pernicious anemia after all, but the cause remained a mystery.

None of these prizewinners discovered the substance in liver that reversed anemia, but they presumed it was a B vitamin lacking in the diet of vegetarians. They were correct: vitamin B_{12} is made only by animals, as it is not needed by plants. Vitamin B_{12} was finally isolated in 1948. It took eight more years before its structure could be determined by Dorothy Hodges in 1956, for which she won the 1964 Nobel Prize in Chemistry.

Vitamin B_{12} deficiency is still a common condition today, but it comes on so gradually that the symptoms may be missed or mistakenly attributed to a mental condition or aging. It is diagnosed by a blood test for B_{12} levels. B_{12} deficiency causes degeneration of the spinal cord and brain shrinkage. Severe deficiency is estimated to affect up to 6 percent of the population, especially in people on restricted diets such as vegans, vegetarians, extreme dieters, and the elderly. Some doctors think that even slightly low B_{12} levels can cause symptoms and estimate that 40 percent of the population has some degree of B_{12} insufficiency.

27

Shades of Cloning

Hans Spemann won the Nobel Prize in 1935 for discovering how the embryo develops. He was born in 1869 and earned degrees in zoology, botany, and physics in 1895. He fell ill in 1896 with tuberculosis, a condition called the "white plague" in those days because it killed up to four hundred out of every hundred thousand in the population. While in bed for his long recovery, he read a book by the evolutionary biologist August Weismann, which inspired him to research embryology.

Following fertilization of the egg, the embryo is the first stage of development of a new organism. The original embryo cells are *generic*, not having the characteristics of any fully developed type of body cell. Spemann was looking for what drove the generic embryo cells to turn into each kind of specialized cell that composes organs, arms, legs, and other body parts. He did various experiments in amphibian eggs using the earliest embryonic stage that was roughly equivalent to the blastocyst in human embryos, which consists of merely two hundred to three hundred cells. Spemann suspected that a region in the cells directed their differentiation.

Spemann initially used the fine baby hair of his infant child, Margrette, to tie a noose around the microscopic glob of salamander embryo cells, trying to see which half contained the organizing cells that directed growth and development. Instead, both halves grew the precursors to a brain and spinal cord, and it developed into a two-headed salamander. This prompted a series of experiments in which different portions of the embryo cell mass were separated by a constriction. The delicate work required the invention of new instruments, so laboratory workers at the Spemann School of the University of Freiburg in Germany constructed delicate glass needles, minuscule glass chambers, and glass bridges, making it possible to manipulate the fine embryonic tissue.

The work was carried out by Spemann's laboratory assistants and graduate students, including Hilde Mangold, who was a PhD candidate and the wife of Spemann's chief laboratory assistant. Working under Spemann, Mangold actually conducted the pivotal experiment that was the basis for Spemann's Nobel Prize. She transplanted some cells from the upper edge of a colorless salamander embryo into a pigmented salamander embryo. The receiving embryo grew into two joined and functioning embryos, one of them having a mix of transplanted cells and recipient embryo cells. She and Spemann concluded that the transplanted cells had an organizing effect on the surrounding cells and directed where they would migrate to and what they would become. They called it the "organizer," located in salamanders at the upper lip of the embryo. This became Mangold's PhD thesis in 1923. The same data were published in a paper in 1924 with Mangold as second author after Spemann. However, Mangold died from burns suffered in a gas heater explosion in her home that year and did not live to see the publication.

This experiment prompted researchers around the world to seek the location of "organizers" in other species. One of the researchers in Spemann's laboratory, Johannes Holtfreter, transplanted embryonic organizer cells from horny frogs into salamanders, which gave rise to salamanders with horns. It was also Holtfreter who was the first to challenge Spemann's concept of an organizing group of cells. Holtfreter's experiments showed that even dead tissue could direct the positioning and differentiation into specific cell types. He concluded that there must be a chemical doing the organizing independent of live cells. Decades more research have yielded the discoveries of a number of different molecules that can act as embryonic inducers to promote cell differentiation. A current popular theory is that a combination of actions is in play, some chemical and some cellular.

Shortly after receiving the Nobel in 1935, Spemann officially retired to professor emeritus status but continued to lecture and persisted in his experimentation. He wanted to determine exactly what was causing the organizing effect now that its location was known. In one experiment, he tried to find out if the organizer effect was in the nucleus of the cell. He transplanted an embryo cell nucleus into a different embryo, in a cell that did not have a nucleus. This was the basis for what is today known as cloning.

All German scientists of the era were affected in some way by the rise to power of the National Socialist Party, and Spemann was no exception.

In particular, Nazi leaders were always on the lookout for any scientific evidence in support of their ideology. Zoologists Otto Koehler and Konrad Lorenz, who were Nazi Party members, publicly cited Spemann's embryonic organizer as a biological parallel to the Nazi principle of *Führerprinzip* (leadership principle). Führerprinzip holds that genetically gifted people are born leaders and the leader's word is law. In an ideal political organization, according to Führerprinzip, each lower division is headed by a leader who similarly holds the power of law over his subordinates. In this scheme, Hitler—the ultimate leader (führer)—was the organizer for the entire world.

Spemann was a nationalist and conservative in his politics but not known to be a member of the Nazi party. He continued at the University of Freiburg, while his Jewish students and research colleagues fled to England and America. His younger son married a Hungarian-Jewish scientist and had to flee. In 1937, two years after becoming a Nobel winner, Spemann publicly asserted that science should be an international effort free of undue national influence. This prompted Nazi leadership to cancel Spemann's right to lecture at Freiberg University. He remained in Germany and died of heart failure in 1941.

28

Brain Chemistry 101

By the turn of the last century, it had been established that nerves have tiny gaps between them, as clearly demonstrated in the drawings of the 1906 Nobel winner Ramón y Cajal. Scientists were hard at work to find out how impulses were transmitted from nerve to nerve. The Nobel Prize of 1936 was shared by Sir Henry Hallett Dale and Otto Loewi for their discovery of the first *neurotransmitter*, a chemical that facilitates the electrical signal to be continued at the next nerve.

Dale was an above-average student whose early school interests were more in the humanities than the sciences. When he was about fourteen years old, he ranked second place in England in school examinations in such subjects as English composition, French, Latin, and scripture. However, the only talk of his future at that time was taking on a clerical job or joining his father in the pottery business. There was no money in the family for college-preparatory school or eventual university.

In fact, the local school was eager to take advantage of its bright student and hatched a scheme to make Dale a first-place winner, all for the pride of the school: it convinced Dale's father to hold the boy back a year and sit for examinations again, which obviously gave him an unfair advantage. This time Dale took courses in economics, bookkeeping, and sciences, and at the end of that year, he did indeed score highest in the country on his repeat exams. This was recognized by various prizes, including one for fifty pounds from Barclays Bank, which was seeking candidates to fill junior banking positions.

Dale's course was suddenly altered when he accompanied his father to a church convention and had a chance meeting with the biblical scholar Dr. W. F. Moulton. Moulton was the headmaster of the relatively new Leys School in Cambridge, a college-preparatory boarding school. The date for entrance examinations had long since passed, but Moulton arranged for tests to be sent to Dale. As expected, he qualified for admission and was awarded a scholarship. At Leys, Dale did well in science, and his teachers pushed him toward courses that would qualify him to compete for a scholarship to University of Cambridge. In the next six years at Cambridge, Dale completed studies in physiology and medicine. He would have preferred to continue in the research laboratories at Cambridge but lost out in the competition for a paid position.

By 1900, Dale had no other option but to return home to work as a doctor at the local hospital. He considered his two years in clinical practice to be a waste since there were no good treatments for most of what ailed people, particularly pneumonia and tuberculosis. He was critical of his fellow practicing doctors, who fell into two distinct camps: one group did not keep informed about the latest medical discoveries and relied entirely on ineffectual remedies based on old wives' tales; the other group pompously claimed (false) scientific bases for its mostly useless modern tinctures.

Dale abandoned clinical practice and found a minor position at the University College London for two years. There he met a visiting research student from Germany, Otto Loewi. They developed a lifelong friendship and would share the Nobel Prize more than three decades later. The short stay in London also made Dale famous for his involvement in the Brown Dog Affair.

The Brown Dog Affair was a controversy about using live animals in experimental surgeries, a procedure termed *vivisection* (from Latin *vivus*, "alive"). Animal rights activists were relatively few in number in England but drew strength from foreigners and other groups who had in common only a general discontent toward the status quo. These included radical Socialists, trade union members, and suffragettes. In addition, the National Anti-Vivisection Society of England was bolstered by the International Anti-Vivisection Council.

The affair was initiated in 1903 by two Swedish animal rights activists who gained legitimate admission as students to the London Medical College for Women, supposedly specifically for the purpose of infiltrating vivisection demonstrations. They attended a *demonstration theater*, a typical operating theater of the era consisting of an operating platform surrounded by tiered seating for medical students to observe. Students watched a demonstration of the hormone research conducted by William Bayliss, who headed the laboratory where Dale worked. Bayliss operated on a small brown dog, a terrier mix of about fifteen pounds. This dog had previously been operated on to clamp a duct going to its pancreas. After the animal was anesthetized, Bayliss cut into the neck and exposed some nerves. He demonstrated that electrical stimulation of nerves in the neck caused the dog's pancreas to produce a newly discovered substance called *secretin*, the first hormone ever named. At the end of the procedure, the dog was sacrificed, a job that fell to the junior member of the team, Henry Dale. He skillfully pierced the dog's heart to bring a quick end to its life.

The Swedish women passed their notes on to the International Anti-Vivisection Council, alleging the dog was not anesthetized and had struggled to get off the table. These notes were read at a council convention by the organization's secretary, the attorney Stephen Coleridge. The story was picked up by the notoriously sensational *Daily Mail*. The newspaper article quoted Coleridge as saying that Bayliss was operating on fully awake animals, and it condemned vivisection for being cruel and unlawful. Bayliss sued for slander and libel. In the lawsuit, Dale testified that he had deftly knifed the animal in a typical surgical maneuver, and the defense made the most of this dramatic method of euthanasia. The judge determined in favor of Bayliss and awarded him court costs and damage fees.

The verdict enraged the activists, who collected funds for building a bronze statue of the little brown dog. They got it placed in a park in 1906 along with a plaque reading: "Men and women of England, how long shall these Things be?" Medical students occasionally vandalized the statue, so it eventually had to be guarded by police. In November 1907, a mob of one hundred medical students from University College London tried to tear down the statue but was warded off by police. Over the next few weeks, so-called anti-doggers protested in disorderly marches across London and were joined by other students and rabble-rousers who probably did not even understand what the ruckus was about. In December 1907, the unruly mob reached Trafalgar Square, where riots broke out complete with fistfights, thrown furniture, and smoke bombs. Protests and scuffles over the brown dog statue continued periodically. In 1910, the local city managers ordered a stealth operation in the middle of the night, in which a few operatives, heavily guarded by police, removed the statue and had it melted down, ending the Brown Dog Affair.

About this time, Dale wanted to marry, but his income did not make him an attractive prospect. Eventually, he took a position at the laboratories of the Wellcome Foundation pharmaceutical company where he would be well paid to conduct physiology experiments of his own, of course with the objective of some discovery that would promote a new drug for Wellcome. He married his first cousin, a practice that was allowed in England but outlawed in much of the United States because having a child with one's first cousin raised the risk of a significant birth defect from about 3 to 4 percent to about 4 to 7 percent. Luckily, the Dales had three normal children: one daughter obtained a PhD in physiology, another earned a bachelor's degree in science, and their son became a medical doctor. The Dales were married for sixty-three years, and they both lived into their nineties.

In his six years at Wellcome, Dale was interested in investigating what made nerves transmit impulses but was diverted to other work that was more in line with products that could be immediately commercialized by Wellcome. He discovered histamine and determined that it caused dilation of blood vessels and contraction of smooth muscles. He saw that histamine was an alarm signal given off by injured cells, which causes nearby capillaries to leak and blood vessels to dilate. He recognized that these were similar to the symptoms of allergic reactions and was later proved correct, but he did not carry that knowledge into drug development. Two decades later, Daniel Bovet discovered drugs that block histamine, which came to be known as *antihistamines*, for which he won the Nobel Prize in 1957.

Toward the end of his time at Wellcome, Dale conducted the initial experiments that led to his Nobel Prize. In 1914, he directed his laboratory chemist to create a chemical called acetylcholine. They tested acetylcholine on animals and found that it had an effect on the nerves by variously exciting, enhancing, delaying, or blocking nerve transmissions. Acetylcholine could affect nerves in the heart, brain, lungs, skin, gastrointestinal tract, muscles, and sweat glands. Dale concluded that chemicals were responsible for the transmission of impulses across the gap between nerves. It was not until Otto Loewi conducted experiments in 1921 that acetylcholine was discovered to be a naturally occurring substance in the body and a natural neurotransmitter.

Otto Loewi was a German born into a family of wealthy wine merchants. He studied classical languages and was planning to go to the university for art history, but his father urged him toward medical school. In his final year, he skipped too many science classes to attend humanities lectures, so he had to remain an additional year before he could complete his medical degree in 1896. He became a physiologist and researcher at the University of Graz, Austria.

Loewi had studied the *vagus* nerves (from the Latin *vagus*, "wandering"). The vagus nerves come out of each side of the lower brain and send branches to the heart, lungs, and digestive tract. When a branch of the vagus nerve is stimulated by an electrical impulse, it causes the heartbeat to slow. But exactly how? Loewi postulated the existence of neurotransmitters as far back as 1903 and struggled with how to prove it. The experiment to test this hypothesis came to him in a dream in 1920. He woke the next morning feeling he had stumbled onto something very important but could not read what he had scrawled on his notepad in the night. The next night the idea returned, and this time he did not want to risk forgetting it in the morning. He arose and went directly to the lab to carry it out.

The famous experiment involved two beating frog hearts, each kept alive in its own bath of artificial solution instead of blood. One heart had its vagus nerve cut, and the other one had an intact vagus nerve. Loewi electrically stimulated the vagus nerve going to the first heart, causing a slowed beat. Then he transferred the liquid solution from the first heart to the second heart, and its beat slowed down too. He also performed a complementary experiment: stimulating a nerve branch that sped up the heart and then transferring the liquid to a second heart, which sped up as well. He concluded that some chemical secreted by a stimulated nerve was causing these effects on heart rate. He suspected it was acetylcholine, but more experiments were needed to confirm that. His work established that the main mode of transmission of nerve impulses was chemical, not electrical.

Dale and Loewi were firmly in the camp of "soupers," who held to the theory of chemical neurotransmitters. There were still some "sparkers," who thought the ultimate transmission was electrical, including their contemporary John Carew Eccles. Eccles won the Nobel in 1963 (shared with two others) for working out the transfer of sodium, potassium, and chloride across the nerve endings' membranes. He described how the ions made an "electrochemical gradient" that caused electrical flow. It was later found that most synaptic signaling is initiated by chemical neurotransmitters, but there are some nerve junctions that are strictly electrical. Eccles's work showed that an event initiated by a neurotransmitter ultimately causes a change in the electrical flow of ions. Thus, both sparkers and soupers were correct.

In 1938, the Anschluss occurred, in which Nazi Germany annexed Austria. In the weeks before Anschluss, there was a massive roundup of political dissenters, including all Jews, Communists, and Social Democrats. An estimated seventy thousand Austrians were imprisoned or sent to concentration camps. Loewi was jailed by the SS for two weeks, but an international uproar of concerned scientists led to his freedom on one condition: he would be allowed to leave the country after transferring his Nobel winnings to a Nazi-designated bank in Germany. His winnings amounted to 5,000 Swedish kronor, today equal to 630 US dollars but then worth 11,000 US dollars. Loewi ultimately found refuge in America and continued his research at the New York University medical school.

The same year that Loewi and Dale received their Nobel Prize, the German chemist Gerhard Schrader was testing various compounds for their strengths as insecticides. He stumbled upon a chemical that inhibited the breakdown of acetylcholine after it had done its job of transmitting the nerve impulse. The drug caused it to persist in its actions, provoking massive overstimulation of the nerves and resulting in death in humans in mere minutes. It was developed as *sarin*, a biological warfare nerve gas eventually possessed by Germany, the United States, and Russia, none of whom are known to have used it. Sarin was thought to be responsible for a 2013 attack on a Syrian suburb, for a 1988 attack on Kurds by Iraqi dictator Saddam Hussein, and for a 1995 terrorist attach on the Tokyo subway.

The discovery of the role of acetylcholine in the body led to the development of medications. It is given as eye drops to cause constriction of the pupil during cataract surgery. Drugs that increase or mimic acetylcholine include nicotine, Chantix, and Alzheimer's drugs. Drugs that decrease or block acetylcholine are the antinausea medication scopolamine, benztropine and biperiden for Parkinson's, Botox injections for wrinkles, and mecamylamine for blood pressure.

29

Starting at the Wrong End

The 1937 prize in medicine went to a Hungarian, Albert Szent-Györgyi, for his discoveries relating to vitamin C. He described himself as a dull young boy and a poor student who had hated books and needed his tutor's help to pass exams. Yet he said as a child he'd learned from his family that the only things worth striving for were the creation of new knowledge or beauty. When he was sixteen years old, Szent-Györgyi spent time with his uncle Mihály Lenhossék, a famous physiologist. Under his tutelage, the young man suddenly developed a keen interest in applying himself earnestly to his studies. He went to medical school but soon became utterly bored again, preferring to spend his time working in his uncle's lab. Uncle Mihály allowed his nephew to be there on the condition that his initial research needed to focus on the anus and rectum—Mihály had hemorrhoids and was hoping for a breakthrough in research that would lead to a cure. Szent-Györgyi later joked that he "started science at the wrong end."

His medical studies were interrupted by World War I when Szent-Györgyi served as a medic in the Austro-Hungarian Army (allied with Germany). He received a medal of valor, but after three years, he was desperate to escape the miserable trenches. His knowledge of anatomy helped him shoot himself in the upper arm to disqualify himself from further front-line service but not disable him. The self-infliction was not detected by his superior officer, so he evaded a court-martial and the usual penalty: execution. Back in Hungary, he finished medical school and was posted in the bacteriology laboratory of the army. He got into trouble when he balked at doing dangerous medical experiments on Italian prisoners of war. As punishment, he was sent to the northern Italian swamps where malaria was rampant and more soldiers of the occupying army died of malaria than gunshot wounds. Luckily, the war collapsed within a few weeks and he was able to return home.

From there he bounced around research laboratories, starting at a pharmacology laboratory in Possony. Within a few months, Possony became part of Czechoslovakia under the conditions of the Treaty of Versailles. During the sudden transition, Szent-Györgyi had to dress as a worker to stealthily get his laboratory equipment out of the heavily guarded university compound. He worked briefly at laboratories in various cities in Germany and then went to the Netherlands. He studied the transfer of electrons in cells (*biological oxidation*) and how cells convert carbohydrates into energy (*intracellular respiration*). This work earned him a Rockefeller Foundation paid-research fellowship position at the University of Cambridge in England. After obtaining his PhD at Cambridge, he dedicated the next fifteen years to studying how cells obtain energy.

Energy from sunlight is trapped by plants, which combine it with carbon dioxide and water to make carbohydrates; oxygen is left over from this reaction. The reverse occurs in animals who eat the plants: the carbohydrates are combined with oxygen to release the energy trapped inside and carbon dioxide is left over. Szent-Györgyi worked out exactly how this happened and found that the cell could not convert a molecule of carbohydrate into energy all in one step. Instead, it took many individual steps to gradually convert a large banknote into small change, as he put it. This was accomplished by the transfer of electrons in the form of hydrogen molecules being passed around and each reaction peeling off bits of energy. Szent-Györgyi described it this way: "The foodstuff, carbohydrate, is essentially a packet of hydrogen, a hydrogen supplier, a hydrogen donor, and the main event during its combustion is the splitting off of hydrogen."[1]

Every reaction has substances that make it go faster (*enzymes*) and substances that act as donated molecules to help the reaction along (*cofactors*). Some of the B vitamins were already known to be cofactors in this cycle. While at Cambridge, Szent-Györgyi found a cofactor that was new, so he had to name it. His first choice of a name was *ignose* (from "ignorant," + *ose*, "sugar"). The editor of the biochemistry journal did not appreciate the joke, so Szent-Györgyi tried *Godnose* ("God knows"), which was also rejected. So he settled on the temporary name *hexuronic acid* (from *hex* "six," for six carbon atoms, + *uronic*, "of the urine").

Szent-Györgyi was next awarded a brief fellowship at the Mayo Clinic in Minnesota, at which he extracted hexuronic acid from animal adrenal

1. Albert Szent-Györgyi, "Oxidation, Energy Transfer, and Vitamins (Nobel Lecture, December 11, 1937)," *Nobel Lectures, Physiology or Medicine, 1922–1941* (Amsterdam: Elsevier, 1965).

glands. From there he went to England, but there were not vast slaughterhouses to supply such glands. He turned to plant sources of hexuronic acid. Szent-Györgyi showed that hexuronic acid was also found in fruits that did not tend to brown when they aged. He next returned to Hungary to be a university professor in Szeged, which was known as the paprika capital in his country. Paprika is made from grinding up dried pods of red bell peppers. Szent-Györgyi noticed that peppers were one of those fruits that did not brown when they aged and correctly guessed that paprika was a rich source of hexuronic acid. Using paprika, he was able to extract enough hexuronic acid to continue his studies. He found that it was the active ingredient in fresh fruit that prevented scurvy and renamed it *ascorbic acid* (from *a*, "against," + *scorbic*, "scurvy"). Ascorbic acid was found to be identical to a mysterious cofactor that had been called "vitamin C" since about 1912, when no one knew exactly what it was. Szent-Györgyi was able to actually isolate vitamin C in 1928. He extracted a few kilos of it from paprika and sent samples to all scientists worldwide who were researching vitamins and cellular energy production. He proposed this simple definition of a vitamin: "A vitamin is a substance that makes you ill if you don't eat it."[2]

Szent-Györgyi was asked by a colleague to send along some pure vitamin C for research and personal use. This person happened to have a disorder that caused him to bleed very easily, a condition called *purpura*. As it was toward the end of the growing season, there was not enough paprika to make a sizable amount of vitamin C, so Szent-Györgyi instead sent him paprikas (dried whole peppers). His colleague was cured of his bleeding problem. They later tried to see if the same results could be obtained with pure vitamin C, but this failed to cause a cure. It provoked Szent-Györgyi to do further research, and he eventually showed that an additional substance makes it possible for vitamin C to work. This was discovered to be a group of compounds called flavones, also known as *bioflavonoids*. Bioflavonoids are found in many plants, especially in parsley, blueberries, black tea leaves, citrus fruits, peanuts, and cocoa. He named these substances vitamin P for "purpura."

Many vitamin manufacturers have attempted to recreate what is found in nature by packaging their ascorbic acid with bioflavonoids in the form of rose hips, pine bark extract, or quercetin. Bioflavonoids have been found to have amazing properties all on their own, including being anti-

2. R. A. Kyle and M. A. Shamp, Albert Szent-Györgyi quoted in "Albert Szent-Györgyi—Nobel Laureate," *Mayo Clinic Proceedings*, 75 (7), 722 Jul 2000.

oxidant, antibacterial, and anti-inflammatory, and to have effectiveness against cancer and hardening of the arteries. They can lower cholesterol, improve blood sugar, and delay the effects of aging. It became feasible to mass-produce ascorbic acid by 1933, but most vitamin makers missed the point and failed to add bioflavonoids to make this process work. Szent-Györgyi traveled widely promoting the many uses of vitamin C combined with bioflavonoids.

In his Nobel lecture, Szent-Györgyi poetically acknowledged his fellows, saying that he had always felt part of a greater international spiritual community. He received the Nobel Prize at the age of forty-four, and it was the biggest lump sum of money he had ever seen. In his book *The Crazy Ape*, he relates, "The easiest way to drop this hot potato was to invest it, to buy shares. I knew World War II was coming and I was afraid that if I had shares which rise in case of war, I would wish for war. So I asked my agent to buy shares which go down in the event of war. This he did. I lost my money and saved my soul."[3] As for his heavy golden Nobel Prize medal, it was worth about six thousand US dollars melted down, which he donated to Finland when the Soviets declared war on that country.

Szent-Györgyi would keep researching for another forty-nine years. He stayed in Hungary through the Nazi era and was very active in the underground resistance movement. He helped many Jewish friends escape to more friendly countries. When on a tour of a government facility, he inquired about jars containing a crude preparation of ascorbic acid. He was told those were for stocking German submarines to prevent scurvy while the men were kept at sea for very long periods. He struggled constantly about the worth of scientific endeavors if the fruits of his labors would only be turned to war efforts.

Although Hungary was allied with Germany, its prime minister was secretly conducting negotiations with Western Allies to help his country break free from the Nazis. Szent-Györgyi offered himself as a negotiator and traveled to Istanbul for secret meetings with British secret service agents under the guise of attending scientific meetings. The secret operation leaked out, forcing Szent-Györgyi into hiding in Hungary. At first, he stayed at the Swedish legation building and had to be smuggled out in the trunk of a car just hours before Nazis ransacked the legation looking for him. He hid with his family in two other locations, which were bombed

---

3. Albert Szent-Györgyi, *The Crazy Ape: Written by a Biologist for the Young* (New York: Philosophical Library, 2007).

within a short time of his leaving them. The king of Sweden secretly had full citizenship granted for his family and arranged for Swedish passports in the name of Mr. and Mrs. Swensen. They never made it to Sweden. By 1944, the Nazis had occupied Hungary, and arrest warrants were issued for Szent-Györgyi and every member of his family, including his young daughter. He spent the last two years of the war in hiding from the Gestapo.

After the war, Hungary became a property of the Soviet Union. Szent-Györgyi was opposed to Communism but stayed on for a couple of years in the hope that scientific freedoms would be protected under the new regime. He even participated in cultural exchange meetings in Moscow but soon saw that mother Russia was cutting ties with all aspects of Western society, including scientific connections. He applied for a temporary American visa to spend a few months at MIT, which he was granted (after his application was held up while the Americans investigated if he was too closely connected with the Soviets). Meanwhile, at home in Hungary, he was being characterized as "a traitor to the people." In 1947, he applied for a visa to relocate to America and again it was delayed while his dealings with the Soviets were scrutinized. In America, Szent-Györgyi was tracked for several years by the FBI and investigated by Joseph McCarthy's House Un-American Activities Committee (HUAC), which never found anything to incriminate him.

He established a laboratory at the Woods Hole Marine Biological Laboratory in Massachusetts with initial funding from a wealthy benefactor and later funding from the Rockefeller Institute and the National Institutes of Health (NIH). His main focus of research was how muscles contract. With additional funding from Armour and Company (in the meatpacking industry), Szent-Györgyi discovered that two proteins, actin and myosin, created high-velocity muscle contractions.

Building on his work with cellular respiration, he turned his attention to cancer research. As far back as 1941, he had hypothesized that cancer was a problem in how electrons moved through tissues. Normal cells used a slow, measured approach to pass electrons in a many-step process, using large molecules to gradually transfer a charge from a carbohydrate molecule into little bits of energy. In contrast, Szent-Györgyi thought that cancer cells had an unregulated approach. He postulated that before oxygen was available on the planet, all primitive cells were in an unregulated growth state that he called *alpha*. As the planet matured and oxygen became more available, complex organisms developed, and they required a *beta* state that suppressed uncontrolled growth. He believed that disruption of the charge-transfer process caused cells to revert to an alpha state and become cancer. In the alpha state, there was no way to mitigate the effect of so-called *free radicals*, meaning unpaired electrons bouncing around. It has since been shown by many others that free radicals can damage cells and accelerate tumor growth, cardiovascular disease, and age-related diseases. Szent-Györgyi thought the salt form of vitamin C, sodium ascorbate, would preserve the beta state, helping to mop up these free radicals. Sodium ascorbate is more bioavailable, and unlike ascorbic acid, it is alkaline, so it does not bother the stomach.

By 1970, Szent-Györgyi's funding had dried up. He refused to apply for grants because the applications always asked how long the research would take and what it would be used for. He retorted that if he already knew what he would do and what he would find, he would not need to do the basic research. In addition, the quantum physics aspects of his cancer research (which described hopping electrons) were too radical for the usual funding institutions. He wrote, "Research is four things: brains with which to think, eyes with which to see, machines with which to measure, and, fourth, money."[4] In 1971, he received a donation of twenty-five dollars from an attorney who read a newspaper account of a speech Szent-Györgyi had given at the National Academy of Sciences. The donor was astonished to receive a gracious thank-you letter from him and proceeded to put together a private nonprofit called National Foundation for Cancer Research. It funded not only Szent-Györgyi but many other researchers in a liberal open-laboratory concept, in which individuals were free to pursue their projects with Szent-Györgyi being the overall intellectual leader.

Szent-Györgyi used electron microscope technology to demonstrate that protein tissue could act like *semiconductors* (a material that partially conducts electric current). Later researchers proved that cancer tumor cells have an electrical conductivity that is different from that of normal cells, as predicted by Szent-Györgyi. He died in 1986. He is still considered worldwide to have been a brilliant scientist. He offered these words of wisdom and encouragement to students of science:

> Think boldly, don't be afraid of making mistakes, don't miss small details, keep your eyes open, and be modest in everything except your aims.[5]

---

4. Albert Szent-Györgyi quoted in *New York Times*, "Albert Szent-Gyorgi Dead; Research Isolated Vitamin C," October 25, 1986.
5. Szent-Györgyi, *The Crazy Ape*.

> Discovery is seeing what everybody else has seen, and thinking what nobody else has thought.[6]

Throughout his life, Szent-Györgyi maintained his social awareness with a world humanitarian viewpoint. He believed that the spirit of science was that of goodwill, mutual respect, and human solidarity. He spoke out against the Vietnam War, against the proliferation of nuclear weapons, and against the military-industrial alliances that promoted war. His philosophy books included *The Crazy Ape*, in which he addressed why "man behaves like a perfect idiot." He wrote that the only hope for humanity would be a new educational system based on a real understanding of moral, aesthetic, and spiritual values. He preached that goodwill and human understanding were the road to peace.

6. 2. I. J. Good, Albert Szent-Györgyi quoted in *The Scientist Speculates* (New York: Basic Books, 1962).

30

Matters of the Heart

Corneille Jean François Heymans was a Belgian physiologist who won the 1938 Nobel Prize for discovering how the nerves played a role in controlling heart rate, blood pressure, and breathing. Corneille studied under his father, Jean François Heymans, a pharmacologist who was rector of the Ghent University in Belgium. The Heymans were partners in much of their research and probably would have been awarded the prize jointly, but the father died a few years before.

It had long been known that when blood pressure increased, breathing was inhibited, and when blood pressure decreased, breathing was stimulated, but no one knew exactly how this occurred. There is a respiratory control center in the brain stem, but that region regulates the urge to breathe on a basic level and does not monitor the moment-to-moment adjustments in breathing and blood pressure.

The Heymans noticed the 1926 work of Fernando de Castro, a Spanish neuroanatomist who drew detailed pictures of the microscopic structure of the nervous system. A protégé of the 1906 Nobel winner Ramón y Cajal, de Castro found and illustrated a minute clump of cells sitting at the fork of the branching of the carotid artery in the neck and named it the *carotid body*. He proposed these cells could measure changes in carbon dioxide and other molecules in the passing bloodstream, and in response, he thought they affected the nearby nerves that in turn altered blood pressure and breathing. Over the next three years he published descriptions of some crude experiments that suggested he was correct, but they lacked the finesse to prove his ideas.

Heymans visited de Castro in Spain, and de Castro visited the Heymans' lab in Belgium, liberally sharing ideas. The Heymans devised a unique experiment in which they laid two anesthetized dogs on their backs. The first dog had his head nearly entirely removed from the body except for

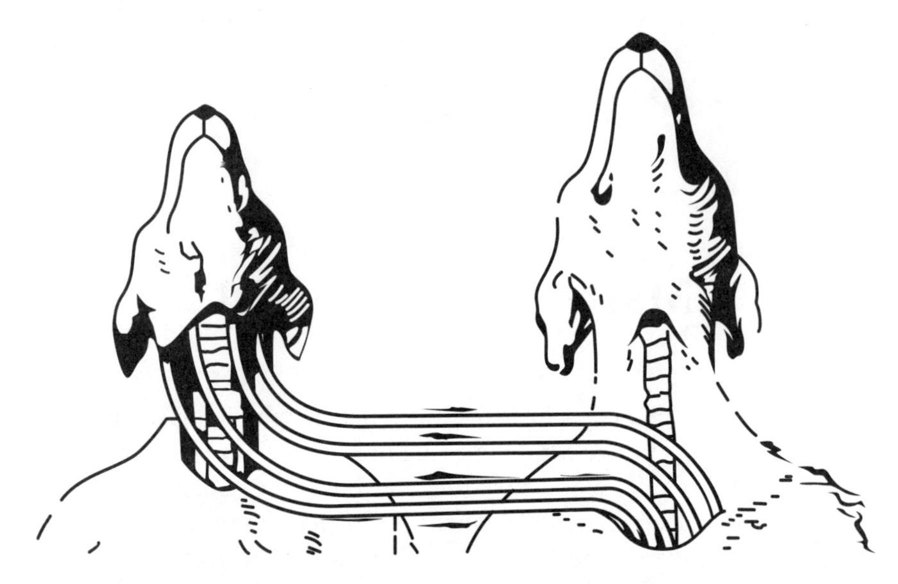

The head of Dog B was only connected to its body by its nerves. The head was kept "alive" by connecting it to blood vessels from Dog A. The Heyman's experiment showed that the expansion of the lungs in Dog A stopped the respiratory movements of the head of Dog B in the expiratory position, which was indicated by the recording of throat and nose movements. Collapse of the lungs of Dog A immediately caused the Dog B's throat and nose to open to take a breath. This showed that there were chemical receptors in the head region of Dog B that were responding to Dog A's release of signaling chemicals into the blood stream.

the nerves that ran between the heart and its head. The second dog had its circulation diverted to deliver blood to the head of the first dog. The Heymans caused low blood pressure in dog 1, and it resulted in faster respiration. Then they caused high blood pressure, and it resulted in slower respiration. This arrangement showed that it must be peripheral nerves rather than the respiratory center in the brain stem that regulated the interaction between blood pressure and breathing and validated the hypothesis of de Castro.

At the same time, two other locations were under study: the *aortic body*, a cluster of cells along the arch of the main artery that leaves the heart; and the *carotid sinus*, a group of nerve endings at the base of the neck artery (a branch of the carotid artery that delivers blood to the brain). It is now known that these other centers do contribute to heart rate, blood pressure, and breathing regulation, but to a lesser extent.

The work of the Heymans was continued by others, and it was discovered that the carotid body is a small cluster of chemical receptors that both detect changes in blood chemistry (especially carbon dioxide levels) and issue chemicals that cause blood vessel walls to dilate.

The time line for Corneille Heymans's Nobel Prize is another instance of how World War II affected scientific affairs. The committee in Sweden responsible for choosing the winner was the Karolinska Institute. It did not decide on a prizewinner in 1938 and did not make its decision in favor of Heymans until the fall of 1939. Then the committee decided to make it retroactive and awarded him the prize of 1938 while it had already deliberated for an official 1939 winner. But 1939 was the eve of World War II. Concentration camps had already been in use in Germany in the 1930s, and Jews and others considered "dissidents" had been rounded up or were disappearing. Many scientists had already fled to England or America. Only one Nobel laureate came to Stockholm to receive his prize in 1939 (the literature winner from Finland). Heymans was presented with the Nobel Prize in his hometown of Ghent, Belgium, in 1940. It was not until 1945 that he delivered his Nobel lecture in Stockholm. He made only a slight reference to de Castro, never crediting him with his original hypothesis and the tremendous contribution of his precise drawings that made Heymans's work possible.

De Castro was never nominated for the prize, but it appears he was at least as qualified for it as Heymans. The Nobel Committee notes released fifty years later showed that someone on the committee had inquired about de Castro; he wondered if de Castro were even alive. A bloody civil war had erupted in Spain in 1936, and the country was split in two: the Loyalist (Republican) side was supported by the Soviet Union while the Nationalist side was supported by Fascist Italy and Nazi Germany. There were atrocities on both sides, including summary executions of anyone caught living in the wrong neighborhood after the split. After Ramon y Cajal there was only one other Spanish scientist to win the Nobel Prize in Medicine in the twentieth century, Severo Ochoa. Ochoa was also a student of Ramon y Cajal, but he had fled Spain at the onset of the civil war and was a US citizen by the time he won in 1959.

31

Brimstone

In 1937, Nazi general Hermann Göring announced a decree that forbade Germans from accepting Nobel Prizes. The regime did not appreciate the international criticisms it received over its brutal treatment of Carl von Ossietzky, the winner of the 1935 Nobel Prize for Peace, awarded in 1936. Ossietzky had been convicted of treason for revealing to the Allied powers that Germany was actively rearming in violation of the Treaty of Versailles that ended World War I and he was then was beaten and starved in a concentration camp. This did not deter the Nobel Committee in Sweden from continuing to award the prize to Germans. Three subsequent winners were forbidden by the Nazis to accept their Nobel Prizes: Richard Kuhn won the chemistry prize in 1938 for his work on vitamins, Adolf Butenandt won the chemistry prize in 1939 for his work on sex hormones, and Gerhard Domagk won the medicine prize in 1939 for his discovery of an antibiotic.

Gerhard Johannes Paul Domagk was born in Lagow, Brandenburg, Germany, in the late 1800s. He decided on a career in medicine when just a teenager, but his first term in medical school was interrupted by World War I. At age nineteen he volunteered for the German Army and was soon wounded, so he was transferred to a medical division. He was assigned to various field hospitals to care for German soldiers, including in Russia, where he saw horrific wounds caused by the new inventions of rapid-repeating rifles and modern grenades. However, by the numbers, what he saw most of was cholera. Cholera is a disease caused by *Vibrio cholerae* bacteria in water that is polluted with sewage. It brings on profound watery diarrhea and loss of electrolytes. It can progress to severe dehydration, shock, and death in a matter of hours when the fluid losses cannot be rapidly replaced by intravenous fluids. More than half a million people were estimated to have died from cholera in Russia in the first quarter of the twentieth century, and there was nothing a field medic could do to treat it.

After the war, Domagk completed his medical degree and went into the research field as an assistant chemist. He wrote a paper on the role of the immune system in fighting infection, which was noticed by recruiters at the Bayer division of parent company IG Farben. The name is a shortened form of Interessen-Gemeinschaft Farbenindustrie Aktiengesellschaft, literally, "community of interest of dye-making corporations." It was formed as a mega-company in 1924 from individual chemical companies that had collaborated in World War I.

The collaboration of the various dye companies included the purpose of agreeing on chemical warfare use and production to support the German effort. The Bayer division was a dye-making and pharmaceutical section that had already synthesized aspirin in 1900. In 1914, at the very beginning of World War I, Bayer CEO Carl Duisberg endorsed the use of chlorine gas as a weapon. This was in direct violation of the Hague Convention of 1899, when Germany was one of twenty-three countries that signed a treaty prohibiting the use of poisonous gases in warfare. After seeing the first test results, Duisberg said, "The enemy won't even know when an area has been sprayed with it and will remain quietly in place until the consequences occur."

Under the auspices of a conglomerate of German chemical companies lead by Bayer, the German Jewish chemist Fritz Haber invented the poisonous gas phosgene, which Duisberg was especially enthusiastic about: "This phosgene is the meanest weapon I know. I strongly recommend that we not let the opportunity of this war pass without also testing gas grenades."[1] Bayer later developed mustard gas. The parent company, IG Farben, would go on to make Zyklon B, which gassed concentration camp prisoners. (Incidentally, Haber won the 1918 Nobel Prize in Chemistry.)

The German Bayer company split off from the American division of the same name, and for many years they were rivals in the worldwide marketing of aspirin. (The two Bayers rejoined as one company in the 1990s.)

It may not seem obvious that a dye-making chemical company would be involved in discovering medicines, but in fact, the very first synthetic dye was accidentally derived from coal tar by a young English chemist who had

1. Carl Duisberg quoted in "History of Chemical Warfare: Poison Gas during World War I: Bayer Still Refuses to Take Responsibility. 100 Years of Chemical Warfare," CBG Network Global Research, April 24, 2018, https://www.globalresearch.ca/poison-gas-during-world-war-i-bayer-still-refuses-to-take-responsibility/5445360.

been trying to make quinine to treat malaria. Since then, the chemical and dye industries have been hand in hand with the pharmaceutical industry. In the 1930s, the IG Farben conglomerate was taken over by the National Socialist (Nazi) state.

Returning to the history of antibiotics: in 1909, researchers in Paul Ehrlich's laboratory discovered that the very first synthetic antibiotic, the dye called 606, later renamed Salvarsan, was found by a systematic testing of hundreds of dye compounds to see which would be effective against the bacteria causing syphilis. Salvarsan became the first "magic bullet"—a man-made drug to treat infection (see chapter 8 for details of this discovery).

After Salvarsan, no progress was made on chemotherapy until Domagk came along. He was hired by Bayer to systematically test its dyes for antibacterial activity. Domagk had a specific interest in the *azo dyes*, which are a group of over 2,000 chemicals composed of nitrogen at the core that are used in industry to color apparel, textile, footwear, leather, and some foods. The term *azo* derives from the Greek *a* "without," + *zoe* "life," as nitrogen does not support life. Domagk was interested in azo dyes because they have a sulfur-containing group in a position that is normally inhabited by a hydrogen atom. Domagk found that such sulfa dyes bound well to the proteins in leather animal hides. He proposed that this property would make sulfa dyes bond to the proteins in bacteria and kill them.

He focused on an orange-red dye named *prontosil*. Even though it did not work when tested on bacteria in the petri dish, Domagk proceeded to test prontosil on infected mice. It worked to cure infection and produced no serious side effects in the treated mice. He withheld publishing his results until he had tested humans and was able to prove that prontosil did effectively treat human infection. One of the human subjects testing the drug was his own young daughter, Hildegarde, who developed a raging streptococcal bacterial infection after an accidental needle prick sustained in her father's lab. She made a full recovery.

Domagk announced his findings in 1935, but it was not until 1936, when the British and American medical societies endorsed the treatment, that some medical centers were willing to try it. A famous case helped to promote the drug's use. In 1936, the American president's son, Franklin Delano Roosevelt Jr., was hospitalized for a sinus infection that spread to his facial skin and went down into his throat: a serious case of strep throat. It was expected to be deadly. Prontosil was suggested to his parents, although the doctors admitted its use was still experimental. They consented to treatment and FDR Jr. got better. Subsequent headlines in the *New York Times* and other newspapers popularized the use of sulfa drugs.

Prontosil was patented by Bayer, but the Nazis never made profits off it. Researchers at the Pasteur Institute quickly discovered that the reason the drug did not work on bacteria in the petri dish but did work in mice and in humans was that prontosil is actually a *pre-drug*. The complete compound is not effective against bacteria. As soon as it is consumed, the body cleaves a section off and only then is the active portion of the drug able to attack bacteria. This active portion is a dye that was synthesized by an Austrian chemist in 1908, but it was unused and forgotten. The original patent on it had long since expired, so anyone was free to make it and modify it for new, patentable drugs.

Chemical companies worldwide set to work inventing variations of the sulfa compound that would work as well or better. While prontosil treated *streptococcus* (strep throat), a variation of the sulfa compound treated pneumonia, another variation treated staphylococcal infection (usually skin infections), and another treated bacterial diarrhea. Initially, the drug was mostly used as a preventive to keep troops on all sides healthy. Sulfa powder became part of a combat soldier's kit, to be sprinkled on any wound as soon as possible. By 1942, the production of sulfa antibiotics in the United States alone was over ten million pounds.

In 1943, while on tour of troops in northern Africa, British prime minister Winston Churchill became ill with pneumonia in Carthage, Tunisia. After some deliberation, he was treated with "M&B 693," a sulfa spin-off created in 1938. He continued to worsen for a week and then made a full recovery. Although penicillin was discovered before prontosil, it was not commercially produced until later and did not even make a contribution to reducing wartime infections until the tail end of World War II.

Sulfa antibiotics indirectly led to much-strengthened US drug regulations. Until then, US consumers were only partially protected by the Pure Food and Drug Act of 1906, which required ingredients to be listed on the label and outlawed the adulteration or mislabeling of a product. What we know today as the FDA was born out of legislation in response to a disaster with sulfa antibiotics.

In 1937, a chemist working at the Massengill pharmaceutical company in Tennessee prepared his own version of a sulfa antibiotic. Harold Watkins used a solvent called diethylene glycol (DEG) instead of alcohol and added raspberry flavoring. It was not widely known that DEG was poisonous and, in fact, Watkins did not know DEG was poisonous, although two professional papers had been published about it. The product underwent tests for appearance, fragrance, and taste, but no safety testing was required. However, at least one report indicated that the company did tests on rats and documented that the mixture caused kidney failure.

The liquid preparation was branded as Elixir Sulfanilamide and distributed in fifteen states. Within a few weeks, the American Medical Association (AMA) received a report of several deaths associated with its use. An extensive operation sought to track down and recall all bottles of the elixir, but there were at least 107 deaths by then, and many of the victims were children.

Massengill's notorious Elixir Sulfanilamide had been mixed with poisonous diethylene glycol, causing more than one hundred deaths in 1937

The newspapers reported outrage and demanded federal action, but the only thing the company was guilty of per existing laws was calling its drug an "elixir," which implied it was made with alcohol as a solvent. The company was given the maximum civil fine of $16,800 for false labeling. The owner and founder of the company, Dr. Samuel Evans Massengill, stated, "My chemists and I deeply regret the fatal results, but there was no error in the manufacture of the product. We have been supplying a legitimate professional demand and not once could have foreseen the unlooked-for results. I do not feel that there was any responsibility on our part."[2] The newspapers dubbed the product "The Death Elixir."

2. Carol Ballantine, Samuel Evans Massengill quoted in "Sulfanilamide Disaster," FDA Consumer Magazine, June 1981, https://www.fda.gov/files/about%20fda/published/

As a direct result of the Massengill disaster, Congress passed the 1938 Food, Drug, and Cosmetic Act requiring that proposed new drugs undergo animal safety tests at company expense and that the results be reviewed by the FDA before marketing approval was given. In 1939, the Massengill chemist, Harold Watkins, committed suicide by shooting himself in the heart.

Domagk was awarded the 1939 Nobel Prize for his discovery of sulfa antibiotics but forbidden from accepting the honor by Nazi decree. After arrest and a one-week detention by the Gestapo, he responded to the Nobel Committee by declining the award. He stayed at IG Farben throughout the war and after it.

At the end of World War II, the big three—US president Harry Truman, British prime minister Winston Churchill, and Soviet premier Joseph Stalin—divided up the spoils of war at the Potsdam Conference. Overnight, Lagow, the town of Domagk's birth, was assigned to Poland. Domagk's mother had been living in the border town of Sommerfeld, which also went to Poland. Many Germans who were stranded within the new Polish borders were treated like prisoners and sent to forced work camps or crowded refugee camps. Domagk's mother died from starvation in a refugee camp.

Domagk ultimately accepted the Nobel medal in 1947 but could not collect the monetary portion because so much time had elapsed. He subsequently researched drugs for tuberculosis and cancer without success.

The-Sulfanilamide-Disaster.pdf.

32

Bleeding Chickens

There were no Nobel Prizes in Medicine awarded in 1940, 1941, or 1942 due to world war activity, but by 1943, the Nobel committees were back in full operation. The medical prize committee at the Karolinska Institute could not agree on a winner in medicine, so they continued deliberating until they chose two winners in 1944 and then retroactively awarded them the prize of 1943. Henrik Dam and Edward Doisy shared the medicine prize for the work they independently conducted on discovery of vitamin K.

Carl Peter Henrik Dam was a Danish researcher who noticed intriguing results in animal research. A study showed that chicks fed on defatted feed tended to bleed. Dam thought the missing ingredient must be cholesterol and conducted an experiment to add back cholesterol. However, he found that chicks have the ability to make their own cholesterol and that adding pure cholesterol to their feed did not reverse the bleeding tendency. He thought maybe the chicks just needed more overall fat, but experiments with adding other oils likewise did not affect the bleeding tendency.

Dam deduced that some other essential ingredient had been in the removed fat. Bleeding was known to be a symptom of scurvy, a disease of vitamin C deficiency. He sought out others' experiences and found a report in which chicks fed on normal fish meal stayed healthy but tended to bleed when they were fed on a specially prepared defatted version. The bleeding tendency was reversed when they were fed cabbage. Those researchers made a presumption that the cabbage supplied missing vitamin C and so did not study it further. The problem was that chicks do not get scurvy (unlike humans, some birds make their own vitamin C and don't need to extract it from their food). Nevertheless, Dam tried adding lemon juice to their diet (a source of vitamin C), and as predicted, this still failed to correct the bleeding.

Next, Dam systematically tried various plant and animal foods, finding several that corrected the bleeding, especially green leafy vegetables and hog liver. Hemp seeds, the seeds of tomatoes, cabbage, soybeans, and alfalfa also worked. He named the missing factor "the coagulation vitamin," or vitamin K for short, because it was a letter of the alphabet that still had not been claimed to name another vitamin. It was also the first letter of the German and Scandinavian versions of the word *koagulation* (blood clotting). He announced his vitamin K findings in 1934.

Subsequently, the American H. J. Almquist was researching the best composition for animal feed lots and discovered that vitamin K is also formed by bacteria in the intestinal canal. In most animals, it is then absorbed into the bloodstream. That failed to happen in chicks because their intestinal tracts were too short to allow for absorption.

In 1939, Dam and his colleagues isolated vitamin K from alfalfa and found it to be a yellow oil in its pure form. However, they still did not know its structure until it was discovered by Edward Doisy later that year. Doisy extracted vitamin K, which is a crystal at room temperature, from liver. Scientists agreed to call the vegetable source K_1 while the animal source is a slightly different molecule called K_2. Vitamin K_3 is a synthetic provitamin that was developed for veterinary use later that same year.

In 1940, Dam was on a lecture tour in America while the German Army launched an invasion of Denmark. It was a six-hour affair with fewer than a couple of dozen losses on either side. Denmark remained occupied by Germany for the duration of the war, so Dam stayed in the United States the entire time. He was a senior research associate at the University of Rochester, New York, when he was notified of his Nobel Prize. He eventually returned to Copenhagen and continued his long career in research at the university there.

Edward Adelbert Doisy was born in Illinois, where he earned his bachelor's and master's degrees. Then he went on to Harvard for his PhD in biochemistry. Most of the remainder of his career was at St. Louis University in Missouri. Before he delved into vitamin K research, he had already discovered the three forms of the sex hormone estrogen: estrone (E1), estradiol (E2), and estriol (E3). These discoveries would seem to warrant a Nobel Prize on their own; in fact, he had been nominated twice in the 1930s.

Another researcher, the German scientist Adolf Butenandt, had independently made a nearly simultaneous discovery about estrone and was awarded the 1939 Nobel Prize in Chemistry for it. A lasting legacy from Doisy's estrogen research was created when he and his fellow researcher at St. Louis University donated the patent for their discovery of the estrogen isolation process to the university. Soon thereafter, Doisy perfected the process and was able to extract a purified version of the hormone from horse urine. It was suitable for human use, and Doisy eventually made financial agreements for its production and distribution by three major pharmaceutical companies.

In 1939, Doisy isolated vitamin K and was able to *synthesize* it (create it in the lab). H. J. Almquist, the same American researcher who found that vitamin K was produced by intestinal bacteria, had also isolated vitamin K around the same time and did receive a single Nobel nomination in 1941. Almquist was a huge contributor to this research, but maybe he was not officially worthy of recognition because he was merely an instructor (not a professor) in the Division of Poultry Husbandry at the University of California's College of Agriculture, not exactly the sort of renowned academic institution typically associated with Nobel nominees.

Doisy had a long subsequent career at St. Louis University School of Medicine, where he made significant improvements in the isolation and identification of insulin. He also studied antibiotics and bile acids.

The use of vitamin K has grown in many more ways than anticipated by its discoverers. It is given to prevent and treat a bleeding disorder that occurs in less than 2 percent of newborns. Affected babies could bleed into their brains, but simply giving vitamin K completely prevents this. It is administered routinely to all newborns, meaning that 98 percent don't actually need it, but the practice seems worth it to prevent a devastating bleed in 2 percent. It is given to people with chronic intestinal illnesses who are not properly absorbing it from foods and those with imbalanced intestinal bacteria. It reverses bleeding tendency in people with liver failure whose livers are not adequately making vitamin K–dependent clotting factors.

Vitamin K is essential for proper bone growth and maintenance of bone health. It promotes the accumulation of collagen, helps bones to attract and keep minerals, and assists in the formation of bone cells, directing which specialized type of bone cell they will mature into. It contributes to the production of *osteocalcin*, a protein that holds calcium in place in the bones. A deficiency of vitamin K probably contributes to osteoporosis and bone fractures.

Along with promoting healthy mineralization of bones, vitamin K is essential for making proteins that prevent soft tissues from becoming abnormally mineralized. It prevents buildup of calcium where it is not supposed to be: in blood vessels, cartilage, skin, and the eyes. The prevention of calcification of arteries is especially important to heart health and brain health. Vitamin K also helps to suppress inflammatory proteins and prevent injury from free radicals, low oxygen, and low blood flow. It has been shown to reduce pain in rheumatoid arthritis and complex regional pain syndrome. Vitamin K cream on the skin promotes the repair of wounds.

Top foods containing vitamin K include green leafy vegetables, *natto* (fermented soy), spring onions, brussels sprouts, cabbage, broccoli, fermented dairy products, prunes, cucumbers, and dried basil. There are three main forms of vitamin K. Vitamin K_1 is from plants (known as *phylloquinone* or *phytonadione*). Vitamin K_2 (*menaquinone*) is from animals. These two are nontoxic and safe to consume. Vitamin K_3 (*menadione*) is a synthetic man-made variant used in veterinary medicine and not recommended for human use. At this time, there is no test to measure the level of vitamin K in the body. However, the natural forms of vitamin K are safe to consume, so it is reasonable to supplement it for treatment or prevention of specific conditions.

33

Hitting a Nerve

The 1944 Nobel was awarded jointly to American physiologists Joseph Erlanger and Herbert Gasser for their work on the functions of single nerve fibers.

Electrical conduction of nerve impulses had been investigated since the late 1700s when Luigi Galvani demonstrated that a dead frog leg could be made to twitch when electroshocked. The first instrument to measure electric current was not devised until a few decades later, and the galvanometer was named after him. In 1803, Galvani's nephew, Giovanni Aldini, conducted experiments on the bodies of executed criminals to see if he could cause any effects. Here, Aldini describes his treatment of the freshly executed corpse of George Foster, who was sentenced not only to death but also to postmortem bodily electrical shocks for drowning his wife and child:

> On the first application of the process to the face, the jaws of the deceased criminal began to quiver, and the adjoining muscles were horribly contorted, and one eye was actually opened. In the subsequent part of the process the right hand was raised and clenched, and the legs and thighs were set in motion.[1]

Legend has it that Aldini's experiments inspired Mary Shelley to write *Frankenstein*. For many years, electroshocking was a fad at parlor games for entertainment and thrills. A few individuals followed the track of Aldini, who was among the first to treat mentally ill patients with shocks to the brain, with horrific results. Meanwhile, mainstream scientists worked at finding out how the body conducted electrical impulses to cause muscle

1. "George Foster Executed at Newgate, 18th of January, 1803, for the Murder of His Wife and Child, by Drowning Them in the Paddington Canal; with a Curious Account of Galvanic Experiments on His Body," *Newgate Calendar*, https://www.exclassics.com/newgate/ng464.htm.

contractions and send messages to the brain about pain and pleasure, among other sensations. Earlier Nobel Prizes in this field included those in 1932 and 1936. Erlanger and Gasser expanded this knowledge further by developing accurate detection, amplification, and recording mechanisms.

Joseph Erlanger was born to California pioneers in 1874. Developing an interest in medicine early on, he was accepted at the University of California with only two years of high school and then went on to Johns Hopkins Medical School. He stayed to teach at Hopkins for a few years, then taught at the University of Wisconsin in Madison before settling at Washington University in St. Louis, Missouri.

He was an enthusiastic mountain hiker, took boxing lessons from former heavyweight champion Jim Corbett, and played the flute in an amateur orchestra. He had a policy to never return to his research lab at night or on Sundays. He ascribed to the tenets of religious humanism, a nontheistic approach to traditional ethical values, and responded to an inquiry about politics on a personal history questionnaire with the word "Mugwump," as in "none of your business." He did not publicly campaign for any particular agenda but when asked about race relations in segregated America, he replied that racism would persist until the color boundaries could be erased by intermarrying, a remarkable public statement in his day.

Before his work on nerves, Erlanger greatly contributed to an understanding of how the heartbeat is related to electrical conduction through specialized fibers in the heart. During World War I, he helped develop a new treatment for shock due to blood loss by composing a blood substitute consisting of sugar water mixed with gum acacia. The gum helped expand the volume of fluid in the bloodstream. Another practical war-related effort was to recommend that war airplanes have their instrument panels positioned at eye level to make it easier to fly blind (without outside visual cues).

Erlanger's former Wisconsin student Herbert Spencer Gasser joined Erlanger's laboratory during World War I. The pair contributed to researching the effects of circulatory shock, a life-threatening medical emergency and one of the most common causes of death for critically ill or injured patients.

Gasser was born in a small town in Wisconsin in 1888 and attained his bachelor's and master's degrees at the University of Wisconsin. He earned a PhD at Johns Hopkins and then returned to the University of Wisconsin for a teaching position. Gasser had a hormonal deficiency that made him look youthful and gave him a high-pitched voice. He strangely wrote his autobiography in the third person.

While Erlanger was working on a blood replacement to treat battlefield wound shock, Gasser was working at American University in Washington, DC, on the development of lewisite, a chemical weapon of mass destruction. Lewisite is an arsenic compound that easily penetrates the skin and latex gloves, causing immediate irritation. It causes chemical burns characterized by water blisters on the eyes, on the skin, and in the respiratory tract. It also causes swollen eyes, a bloody nose, shortness of breath, vomiting, diarrhea, and low blood pressure leading to death from shock. Lewisite was ultimately manufactured and stockpiled in the United States, Germany, the Soviet Union, and Japan, but no country admits to ever using it. All known US stockpiles were destroyed as of 2012.

In 1921, Erlanger brought Gasser into the Physiology Department at Washington University. Their pivotal experiments were conducted in 1922. The existing instruments for studying nerves were neither fast nor sensitive and could not make accurate recordings. A new technology introduced by the Western Electric Company was the cathode ray tube (CRT) for commercial application in television sets. The cathode ray tube allowed for the movement of electrons in an almost interference-free vacuum tube, which would be perfect for detecting the minute impulses of single nerve fibers.

When a CRT is hooked up to an oscilloscope, the tracings of nerve transmissions could be amplified and recorded. Erlanger asked Western Electric for a CRT for their experiments but was refused. Erlanger and Gasser proceeded to construct their own contraption using the same principles. Even though it was awkward and clumsy, their homemade cathode ray oscilloscope did indeed accurately detect, amplify, and record single nerve firings. They succeeded in amplifying a nerve impulse by two million times, which made an easy-to-read deflection on a moving roll of graph paper.

They discovered that nerve fibers of different sizes have different conduction velocities. The thickest fibers pass an electrical signal at five to one hundred meters per second for a maximum of almost thirteen thousand miles per minute. The thinnest fibers conduct an impulse of less than two meters per second. In between are intermediate thickness fibers that conduct at three to fourteen meters per second. They also showed that pain signals mostly travel on the slow conducting fibers while signals for muscle movement and touch perception travel on the fast fibers.

Gasser had been nominated for the Nobel Prize on eighteen occasions starting in 1935, while Erlanger was nominated seventeen times. Gasser's winning nomination was submitted by Professor Evarts A. Graham, the head of surgery at the Washington University, who had nothing to do with the physiology laboratory. It is suspected that Graham's wife, Helen, suggested the nomination. Helen Graham was a neurophysiologist in the Pharmacology Department at St. Louis, where she had worked closely with Gasser. It had been Gasser who broke the conventional prejudice against promoting women by appointing her to that post. Her position was not lofty enough to qualify as a Nobel nominator, but her husband's position was. When they were first nominated in 1938, Erlanger asked Gasser to submit additional information about his work to the Nobel Committee at the Karolinska Institute. Gasser refused, considering it shameless self-promotion.

By the time Gasser was notified of winning in 1944, he had long since forgotten this research and had to go back to his early papers and brush up on the topic of single nerve fibers to prepare for his Nobel lecture.

In the 1930s, Gasser took a position at Cornell University, and from 1935 to 1953, he directed the Rockefeller Institute for Medical Research. There Gasser managed multiple contracts with the Office of Scientific Research and Development (OSRD), a US government office for coordination of scientific work to support war efforts. The OSRD had almost unlimited funding and resources. In addition to the development of chemical weapons, it worked on building more deadly conventional bombs and hand weapons and ran the Manhattan Project, which created the H-bomb that was dropped on Nagasaki and Hiroshima. The OSRD used conscientious objectors in Civilian Public Service (CPS) in human experimentation from 1943 to 1946, the extent of which is not known. Civilian Public Service was a US government program that provided conscientious objectors with an alternative to military service during World War II.

Erlanger stayed put at Washington University, where his clarity of science teaching and communication skills helped attract other talented researchers and elevate the university to international prominence.

There were no Nobel Prize festivities in 1944 in Stockholm due to the ongoing war. Instead, a luncheon was held at the Waldorf Astoria Hotel in New York, organized by the American Scandinavian Foundation. Only some of the 1943 and 1944 laureates, including Herbert Gasser, were able to attend and receive their prizes from the Swedish minister, who was the chief diplomat in Washington.

34

The Rediscovery of Penicillin

Alexander Fleming, Howard Walter Florey, and Ernst Boris Chain shared the 1945 Nobel Prize for their discoveries on penicillin, although penicillin's story dates to well before the work of these Nobel laureates.

In 1897, a medical student and doctoral candidate, Ernest Duchesne, was attempting to earn his degree at the Pasteur Institute in France. He studied the long-known observation that some bacteria and molds are antagonistic to each other's survival. He found plenty of examples of bacteria winning out over mold but sought to find if it ever worked in reverse. Did mold ever inhibit bacterial growth?

He cultured the mold *Penicillium glaucum*, which is what makes the distinctive stink of Gorgonzola, Stilton, and other blue cheeses, and found it did kill off some bacteria in the testing dish. He next tested it in guinea pigs infected with bacteria and found they survived their illnesses if injected with the juice from *P. glaucum* mold. He submitted a thesis proposing the antibacterial activity of *Penicillium* mold, but the Pasteur Institute did not accept it, and his paper was forgotten for the next two decades. Duchesne went on to earn his medical degree after an interruption by enlistment in the army. Eventually, he fell ill with tuberculosis and died in obscurity in a TB sanatorium at the age of thirty-seven.

Costa Rican scientist Clodomiro Picado was but ten years old when Duchesne was writing his thesis. Picado was a bright student who earned a scholarship to study in France. It is unknown if he ever discovered Duchesne's earlier paper. After earning his doctor of science degree, he joined the research staff at the Colonial Pasteur Institute in Central America. His surviving papers included a 1923 article in which Picado documented the antibacterial effect of *Penicillium* molds and a 1927 paper on its use against actual infections. His subsequent work focused on snake-bites and development of antivenom.

Picado is also remembered for being an ardent racist, raising the alarm in his country about the "contamination" of pure European-stock blood with that of African Americans. He was so moved by this trend that he wrote in a letter: "OUR BLOOD IS BLACKENING! If we continue like this, it will not be a nugget of gold that comes out of the crucible, but rather a piece of charcoal"[1] (all capital letters in the original).

It is not known if Alexander Fleming was aware of the earlier papers by Duchesne and Picado, since Fleming's discovery seems to have been purely accidental. He was a respected microbiologist at St. Mary's Hospital in London but also known for having a notoriously messy laboratory. In 1928, he was in the midst of research on staphylococcus bacteria when it came time to leave for summer vacation in August, and he simply stacked all of his culture plates growing the staphylococci bacteria on a corner bench. Upon his return in September, he found that one plate had been contaminated with a mold growth. His laboratory was one floor above the laboratory belonging to a mold expert who was growing *Penicillium notatum*, so probably billions of mold spores were floating around in the air. Remarkably, the culture plate showed a clear zone around the mold in which no staphylococci bacterial colonies were growing. He correctly recognized that the *Penicillium* mold contaminating his culture plate had antibacterial action. He is famous for reporting that his first thought was, "That's funny!"

Next, Fleming intentionally grew cultures of the *Penicillium* mold and tested its antibacterial activity against several different disease-causing organisms. The mold worked against the bugs causing scarlet fever, pneumonia, diphtheria, gonorrhea, and meningitis. His report of these remarkable findings in an obscure journal got little notice in 1929. He continued to work with the mold, which was difficult to grow as a pure uncontaminated culture, and struggled with isolating from it the actual component that had antibacterial action. The laboratory at St. Mary's did not have the resources for the large-scale work required to more efficiently grow the mold or conduct experiments to isolate its active component, and Fleming failed to interest skilled chemists or pharmaceutical firms in this work. The mold juice seemed to act slowly over several days, so he concluded it would not last long enough in the body to do an effective job of fighting human infections. By 1940, Fleming abandoned the project.

1. Clodomiro Picado, "Our Blood Is Blackening," (1939) in *The Costa Rica Reader: History, Culture, Politics*, eds. Steven Palmer and Iv‡n Molina (Durham, NC: Duke University Press, 2004), 243-44.

Howard Florey was an Australian pathologist. His research interest at Oxford University in England was the same that had occupied Duchesne in France: the contest that exists in nature in which some molds kill bacteria and some bacteria attack molds. In 1938, Florey's junior researcher, Ernst Chain, came across the ten-year-old article by Fleming. Florey immediately assembled a team to see if it could solve the problems of *Penicillium*'s growth and the isolation of the magic medicine contained in its juices.

Dr. Ernst Chain was a German biochemist born in Berlin but had three strikes against him in Hitler's Germany: he had Russian and Jewish ancestry and was sympathetic to leftist political principles. Chain immigrated to England in 1933, leaving well before many others at risk recognized the impending threats to their freedom. He was not successful in getting his mother and sister out of Germany, and they perished in the Holocaust.

Florey's Oxford team made a crude preparation of penicillin and in 1940 used it to treat infected lab mice. Mice who received penicillin survived their infections, while all the mice that did not get treated died. Florey and Chain estimated that it would take over five hundred gallons of mold culture fluid to get enough pure penicillin for treating just one infected human patient.

Dr. Norman Heatley, a biochemist, was the third man on the Oxford team. Heatley devised a more efficient way of growing *Penicillium* mold in porcelain hospital bedpans and then expanded on his success by having a pottery shop manufacture bigger porcelain vats for raising crops of mold. However, the resulting penicillin production was still less than what was needed.

Their first patient was a local police constable who had wounded his face while gardening and developed a severe infection extending to his eye, shoulder, and lungs. He had been in the hospital for several weeks under treatment with sulfa antibiotic, but the infections continued to worsen. Permission for emergency use of the penicillin extract was granted, and after five days of injections, there was some improvement. Unfortunately, there was insufficient penicillin to continue treatment. They even re-extracted the penicillin that came out through the patient's urine, purified it, and gave it back to him. The constable ultimately died of his infections a month later.

While England was at war with Germany, Heatley and Florey traveled in a blacked-out plane to a laboratory in the United States in 1941. They reached the Department of Agriculture in Peoria, Illinois, where they collaborated with Dr. Andrew. J. Moyer on developing industrial-scale production methods. They fed the mold on *lactose* (milk sugar) instead of glucose. Moyer added the idea of continually shaking the flasks, which further increased the mold growth and sped up production.

They had the idea of growing the *Penicillium* mold in a liquid broth that contained corn-steeped liquor as food for the mold. Corn liquor was a by-product of the huge grain industry in Illinois and the surrounding Midwestern states. It is the thick fluid left behind after steeping corn kernels, which softens them up to more easily mill them in the production of livestock feed.

After some weeks of close collaboration, Heatley noticed Moyer was becoming more secretive about the results of their experiments. It turns out Moyer was thinking ahead. He gave no credit to Heatley when it came time to publish a paper on the large-scale production methods, and he cornered the patent rights for himself.

Meanwhile, Florey and Heatley were searching for a strain of *Penicillium* mold that would be more productive of the valuable antibiotic juice. Mary Hunt was a United States Department of Agriculture (USDA) laboratory assistant tasked with shopping the markets for moldy breads, fruits, and vegetables, earning her the nickname Moldy Mary. One day she brought in a cantaloupe covered with a golden mold, which turned out to be a species called *Penicillium chrysogeum.* It yielded two hundred times the penicillin as the *Penicillium notatum* mold. They tried treating it with various agents to make it even more productive and finally struck upon a combination of hitting it with X-rays and filtering it to yield a thousand times the penicillin as the original batches of Fleming's mold.

Commercial production of penicillin was suddenly financially attractive. The new production methods combined with the United States Army lining up as the first large-scale customer lured twenty large pharmaceutical firms to enter into penicillin production. In the first half of 1942, 400 billion units of penicillin were made; by 1945, US companies alone cranked out 600 billion units per month. Penicillin production was also going full tilt in Japan, the Soviet Union, and Germany.

All the while, researchers were in the dark about the actual composition of penicillin. In 1945, Dorothy Hodgkin, another Oxford researcher, determined penicillin's structure, which ultimately contributed to her winning the 1964 Nobel Prize in Chemistry. Subsequently, it was found that penicillin worked by inhibiting the reactions needed for bacteria to construct their cell walls.

There were signs of antibiotic resistance very soon after penicillin was first put into use in 1938. As early as 1940, it was detected that some bacteria were capable of producing an enzyme that inactivated penicillin. By 1945, the year of the prize, some people were infected with bacteria that were totally resistant to penicillin. In his Nobel acceptance speech, Fleming himself warned of a time when penicillin would become useless due to irresponsible overuse.

To combat resistance, drugmakers have combined semisynthetic penicillin (*amoxicillin*) with the chemical *clavulanate*, which in turn tricks the inactivating enzyme into binding with it instead of with penicillin. While clavulanate is busy neutralizing the enzyme, the amoxicillin is free to go to work killing the bacteria. But even clavulanate does not fully address bacterial resistance. Resistant organisms used to occur only in people who had been seriously ill with complex conditions and on antibiotics in the hospital for a prolonged time. Within a few decades, the resistance to penicillin derivatives has become widespread, turning what used to be simple conditions into runaway infections that can disfigure and kill, totally unresponsive to the usual drugs. The World Health Organization (WHO) has listed global antibiotic resistance as a public health emergency since 2013.

The Nobel Prize was awarded to Fleming, Florey, and Chain, but not to Heatley, Morley, Picado, or Moldy Mary. In 1990, Oxford honored the contributions of seventy-nine-year-old Heatley when it awarded the biochemist with an honorary doctorate of medicine, the first such award to be bestowed upon a nonphysician in Oxford history.

Although Alfred Nobel stipulated in his will that the prizes should be awarded to those who "conferred the greatest benefit on mankind," Florey had no such humanitarian goal. He said:

> People sometimes think that I and the others worked on penicillin because we were interested in suffering humanity. I don't think it ever crossed our minds about suffering humanity. This was an interesting scientific exercise, and because it was of some use in medicine is very gratifying, but this was not the reason that we started working on it.[2]

In fact, Florey feared that his scientific contributions inadvertently contributed to what he considered an even greater threat: overpopulation due to improving medical care. He was obsessed with the problem and led the Royal Society's Population Study Group until his death in 1968.

2. Hazel de Berg, interview with Sir Howard Florey, quoted in Denise Sutherland, "Sir Howard Florey—A Driven Spirit," *Australasian Science* (February 1998): 9, http://www.asap.unimelb.edu.au/bsparcs/exhib/journal/as_florey.htm.

In the late 1950s, Ernst Chain's research turned toward *lysergic acid diethylamide*—the potent hallucinogenic drug better known as LSD, a synthetic creation derived from the mold that turned rye crops into sprouting, poisonous kernels. Eating the moldy grain resulted in a dangerous clinical syndrome called *ergotism*, characterized by hallucinations, seizures, coma, and death. LSD was first synthetically produced in 1938 by Dr. Albert Hofmann at Sandoz Laboratories in Switzerland. Chain worked on improving ways to produce LSD at Oxford and would continue to publish on this research through 1970. In a 1964 article, he boasted that his lab had so improved methods of LSD production and extraction that it "approaches what was reached in the microbial production of penicillin." He also developed new synthetic strains of LSD.

In 1967, Chain testified in a notorious case defending two English LSD dealers, John Esam and Russell Page. He testified that LSD did not meet England's definition of a poison at the time. On the prosecution side was Dr. Albert Hofmann, the original creator of LSD. Hofmann gave the opinion that LSD was most definitely a poison. The jury acquitted the defendants.

Fleming did not have much to do with penicillin after 1940, although he collected many honorary degrees and distinguished academic recognitions. He was even named Chief Doy-Gei-Tuan (Maker of Great Medicine) by the American Indian Kiowa tribe. He died in 1955.

35

Godzilla in the Making

Hermann Joseph Muller was awarded the 1946 Nobel Prize for his discovery that X-ray irradiation causes genetic mutations. His nomination and election for the prize were all the more remarkable because they recognized Muller's phenomenal scientific contributions while ignoring the complicated personal, professional, and political controversies that dogged him.

Born in 1890 in New York City, Muller attended public school and was a science enthusiast early on, forming a science club with his high school mates. At the age of sixteen, he excelled on his college entrance exams for which he earned a scholarship toward a college of his choice. He enrolled in Columbia University in New York the following year.

Muller stayed at Columbia for graduate studies, working on genetic mutations in fruit flies in the lab of Thomas Morgan. Muller and Morgan coauthored a book on genetic mutations that illustrated the way chromosomes randomly swap genetic information with each other during the development of egg cells and sperm cells. This came to be known as *crossing over*, which describes how genetic material from one chromosome crossed over to another while the original chromosome may get a snippet of genetic material in exchange. This is one way new mutations develop, and they are often dangerous to the body; some are even lethal. Muller earned his PhD with his thesis on crossing over, and Thomas Morgan won the 1933 Nobel Prize for his leading role in this genetic research.

In 1916, Muller moved himself and his flies to continue his genetics research as a faculty member at the newly formed Rice Institute in Houston, Texas. (It would become Rice University in 1960.) In 1920, he moved again with his flies, this time to the University of Texas at Austin. There he influenced colleagues to abandon their work on armadillos and spiders and to adopt the fruit fly as a much easier live laboratory specimen to work with.

This created rivalry as their gene experiments on fruit flies began to overshadow Muller's own work on the effects of X-rays on genetics.

Muller ran into trouble in analysis of the immense amount of data he was generating by mating fruit flies and tracking mutations through the generations. He enlisted the help of Jessie Jacobs, PhD, an instructor in the Mathematics Department. She constructed the mathematical models used in his genetic research, and they coauthored a paper together describing the mathematical prediction of mutations. They soon married, which was a second strike against Muller at UT Austin, where personal relationships among staff were discouraged. When Jessie became pregnant, she was fired due to a sentiment at the university that academia and motherhood were incompatible, a common reaction in the 1920s.

In Austin, Muller conducted his prizewinning experiments on the effects of X-rays on genetic material. The offspring of flies exposed to radiation had significantly more mutations than the offspring of flies not exposed. He found that X-rayed *germ cells* (which produce sperm and eggs) had 150 times the frequency of crossover mutations compared to untreated germ cells. He proposed that heat generated at an ultramicroscopic level by X-rays caused a random movement of molecules, giving rise to accidental crossing over. This fundamentally changed the gene, and sometimes the resulting genetic change was lethal to the organism. Muller published this work in 1927. Other researchers refuted his conclusions and instead proposed that X-rays caused only rearrangements on the chromosome rather than specific gene mutations.

In addition to the academic rivalry within and outside UT Austin and his extracurricular activities with a fellow faculty member, Muller's political pursuits were a source of embarrassment for the university. He served as an advisor to the National Student League (NSL), which had been named by the FBI as a Communist organization. This is likely when Muller first became a "person of interest" to the agency. There are no details, but his biographers make reference to a period of intense FBI harassment. Muller edited a newspaper distributed on campus called *The Spark*, which promoted Communist causes. The local Austin paper attacked Muller as a Communist, although he never joined the party.

In the midst of all of these controversies, relations became strained between Muller and his wife. One night in 1932, Muller took an overdose of barbiturates and did not show up for class the next day. His wife organized a search posse, and he was found wandering in the woods in the outskirts of Austin, disheveled in appearance and in a muddled state of mind. He was back in class the next day, and nothing more was said of it.

Shortly thereafter, Muller further provoked professional disdain by presenting a paper at the Third International Congress of Eugenicists held in New York that year. He denounced the American genetics movement, which had much in common with German eugenics principles. Muller asserted that there was no scientific support for a genetic basis of poverty, feeblemindedness, or habitual criminality and called for an end to discrimination based on class, gender, or race. He also did not agree with so-called *negative eugenics*, which advocated the sterilization or murder of people with undesirable traits. He promoted the idea of *positive eugenics*, in which a truly equal society (such as in the Soviet Union, he thought) could support selection of persons with the fittest characteristics for optimum reproduction.

Later that year, Muller took a leave of absence from the University of Texas in order to take advantage of a Guggenheim Fellowship as a visiting professor to the Brain Institute, part of the Kaiser Wilhelm Institute in Berlin. But that was also the very year that Hitler was elected to lead National Socialist Germany. The Nazis vandalized Muller's laboratory as part of a program of systematic harassment of any suspected Communists. Muller did not stick around, heading east to the Soviet Union, where he was invited to establish genetics laboratories first in Leningrad and then in Moscow. While he was there, his marriage ended in divorce in 1935.

Still holding to the idea that a Communist state would be an ideal setting for his eugenics utopia, Muller expanded his ideas in a 1936 book titled *Out of the Night: A Biologist's View of the Future*. He advocated that reproduction should be scientifically controlled and entirely separate from the conventions of love, sex, and marriage. Muller described a society in which women would be free to line up to receive insemination by those males chosen for their vigor and fitness, high intelligence, and superior social characteristics. Muller's examples of such ideal men included Lenin, Marx, Pushkin, Newton, Da Vinci, Pasteur, and Beethoven. It also included non-Caucasians such as Chinese nationalist Sun Yat Sen and Persian mathematician Omar Khayyam. Muller had his book translated and presented it along with a lengthy letter to the Soviet premier, Stalin. It was a terrible miscalculation for at that moment Stalin was embracing the genetic theories of an obscure Russian crop scientist, Trofim Lysenko.

Lysenkoism proposed that genetic factors could be overcome by the influence of environmental stresses. For example, summer wheat exposed to cold could be made to grow like winter wheat. Lysenko led gullible people to believe that environmental factors could even cause the wheat to turn into rye or oats. Muller publicly debated Lysenko in 1936, calling him a fraud and his work "shamanism." But Lysenko and his followers gained the support of the state, and Stalin adopted Lysenkoism as the national genetic policy. Eventually, anyone who spoke critically of Lysenkoism was at risk of personal harm or even death. Some three thousand biologists were fired, imprisoned, or executed for being merely affiliated with classical genetics or daring to speak up against the sham science of Lysenkoism.

Stalin was displeased by Muller's book and ordered an attack on the book and on persons advocating its scientific tenets, as these were in direct opposition to the state-sanctioned dogma of Lysenko. For example, the secretary who had translated the book into Russian was later arrested and shot for treason.

Muller kept his mouth shut, not only for his own safety, but also out of concern for his colleagues at the laboratories he had set up. He had to find a way out that was politically acceptable. His solution was to volunteer for the International Brigade on the side of the Revolutionary Army in the Spanish Civil War, aligning with the Communists and anarchists against General Franco's Fascists. In this way, he was allowed to exit mother Russia. In Spain, he held out through the siege of Madrid and, when the Revolutionary Army was on the verge of collapse, Muller used his contacts to secure a temporary position at the University of Edinburgh in Scotland.

In experiments in Edinburgh, Muller demonstrated that a single dose of X-rays given over a period of thirty minutes caused the same increased rate of mutations as did the same total dose when given slowly over a protracted period of thirty days. He concluded that X-rays were not safe at any dose and raised the alarm in the medical field. He called for technicians and doctors to wear protective lead shields and urged physicians to be conservative about exposing patients to repeated X-rays. This was not well received by medical doctors who heavily criticized Muller as only a fly expert who never had anything to do with patients. It did provoke greater awareness, but it was many years before hospitals implemented meaningful monitoring of radiation exposures. The call to protect patients was not heeded until very recently. Routine mammography for breast cancer surveillance has been documented to be associated with a small but definite excess of naturally occurring breast cancers. Excessive use of CT scans in childhood has now been recognized as tripling the risk of leukemia and brain cancer.

Muller's work in Edinburgh was suddenly halted by the outbreak of World War II and the Luftwaffe bombing London in September 1940. It was strongly suggested to Muller that he return home. At this point, he was still under FBI investigation and not welcome anywhere. The Russians accused him of being an American spy, while the FBI suspected he was working for the Russians, and the Fascists had also attacked him for his stance in Spain. He was informed that if he wanted to become eligible to rejoin the University of Texas at Austin, he would have to stand trial in the university senate for violating a policy that prohibited faculty from editing unauthorized newspapers.

Muller managed to find a temporary slot at Amherst College in Massachusetts. In his short stint there, he studied genetic mutations that occur spontaneously with aging, independent of exposure to X-rays. He and future Nobel winner Barbara McClintock had recognized a tendency for aging chromosomes to lose their end parts, causing them to be unstable. They thought this region was protective of stability, and when lost, the chromosome was more prone to damage. Muller coined the term *telomere* to describe this terminal part of the chromosome.

It wasn't until the late 1970s that definitive research on telomeres proved they were indeed protective of chromosomes, for which three researchers received the Nobel Prize in 2009. Today, telomeres are known about in detail. Telomere shortening limits life span and is associated with age-related degenerative diseases. While at Amherst, Muller was also consulted on the Manhattan Project—the secret operation to make an atomic bomb—and he advised the government on the risks of radiation-induced genetic mutations.

As the brief stint at Amherst was winding down, Muller was desperate to find a stable research position. His letters to friends and colleagues expressed the depths of his despair. In 1945, he was at last invited to join the zoology faculty at Indiana University in Bloomington, for which he was very grateful. He stayed in Indiana for the remainder of his career but continued to provoke controversy.

When Muller was awarded the Nobel Prize in 1946, only a year had passed since the atomic bombings of Hiroshima and Nagasaki and a few months since the US military conducted Operation Crossroads at Bikini

Operation Crossroads, "Baker Day" Underwater Atomic Bomb Test, Bikini Atoll, July 25, 1946

Atoll, testing even more powerful atomic bombs. Operation Crossroads resulted in radioactive contamination of service members, observers, islanders, and cleanup personnel. In fact, the radiation contamination was so extensive that the official cleanup plan had to be abandoned because the efforts were not effective. The islands were belatedly evacuated and remain uninhabitable to this day.

These events served to awaken the attention of the general public to concerns about the mutation effects of radiation, which Muller had been warning about for so long. Muller and other concerned scientists became more vocal in scientific meetings as well as in the lay press. Their antinuclear messages were varied but had in common a central concern for the impossibility of avoiding global damage in an effort to selectively use atomic weapons.

In 1953, Muller was issued a subpoena by two FBI agents to appear before the House Un-American Activities Committee (HUAC). He testified that after his chilling experiences in the Soviet Union, he had adopted the attitude of "better dead than Red," but the FBI continued to keep an eye on him.

In 1954, the US military conducted another notorious test called Castle Bravo, also in the Bikini Atoll. The blast was two-and-a-half times more powerful than predicted at fifteen megatons, one thousand times the explosive power of the Hiroshima bomb. The extensive fallout necessitated another round of belated evacuations of natives living on farther-flung islands to the east of the bomb site, many of whom were already in the throes of acute radiation sickness by the time evacuation teams arrived. The radioactive fallout spread around the globe.

The Castle Bravo bomb generated radiation that reached the crew of a Japanese tuna-fishing vessel named the *Lucky Dragon*, which was approximately eighty miles from the test site. The crew of twenty-three suffered acute radiation sickness, and their captain was the first to die. His funeral was attended by over four hundred thousand in Japan, part of a nationwide protest against the use of nuclear weapons. That same year, the very first Godzilla movie was a reaction to the realities of radiation-induced genetic mutations. The original Japanese version, *Gojira*, was a frightening portrayal of the revenge of nature on humans for developing the bomb. The movie opened on a peaceful scene of the decks of a fishing boat suddenly disrupted by a blinding blast, a boiling ocean, and men attempting to flee in terror. The monster was a genetic freak caused by an atomic bomb. At the end of the movie, after Godzilla was finally brought down, a wise old paleontologist said, "If we keep conducting nuclear tests, another Godzilla may appear somewhere in the world."

In 1955, Muller was invited to present a paper at the United Nations conference on the peaceful use of atomic energy. The US Atomic Energy Commission disallowed his participation and withdrew his planned presentation, "The Genetic Damage Produced by Radiation," from the program. The Atomic Energy Commission was influenced in its decision by the thick files on Muller shown to them by the FBI. Like many antinuclear activists, he had been branded a Communist sympathizer, and his efforts to warn of the dangers of atomic technology were characterized as a ruse to get the United States to stand down in the Cold War while Russia secretly built up its nuclear program.

That same year, Muller and nine other Nobel laureates signed the Russell-Einstein Manifesto along with Polish physicist Leopold Infeld. Albert Einstein and Bertrand Russell initially drafted the manifesto, calling for a nonpolitical meeting of world scientists to discuss the dangers of weapons of mass destruction. A quote from the manifesto read, "Remember your humanity, and forget the rest."[1] In 1958, Muller and many other prominent scientists signed a petition to the United Nations that was initiated by

1. *Russell-Einstein Manifesto*, Atomic Heritage Foundation, atomicheritage.org/key-documents/russell-einstein-manifesto.

another future Nobel winner, Linus Pauling. It called for an end to nuclear weapons testing.

Muller's Nobel Prize changed the attitude of the University of Texas at Austin. When the former department head sent Muller a request in 1954 for a good photo to place in the UT Hall of Fame, Muller replied, "I was glad to know that at last I was to be hung in Texas and not hanged there."[2] Muller died of a heart ailment in April 1967 in Indianapolis, Indiana, at the age of seventy-six.

---

2. H. J. Muller to Carl Hartman, October 2, 1954. Lilly Library, Muller Archives, Indiana University, Bloomington, Indiana.

36

Sugar Metabolism

The 1947 Nobel Prize was awarded for research relating to the metabolism of sugar (glucose) and shared: half went to the Argentine physiologist Bernardo Houssay and the other half was split between the naturalized American married couple Carl Cori and Gerty Cori. It was the first prize in the field of medicine to be awarded to a South American national, to a woman, and to an American married couple.

Bernardo Alberto Houssay was born in Buenos Aires, Argentina, in 1887 and received his primary education at the English-language British School, where he was a phenomenal young student, completing the entire curriculum at the age of seven. He was admitted to the University of Buenos Aires Pharmacy School at only fourteen years of age, went to medical school at age seventeen, and was made a full professor in the physiology department a year before he graduated from medical school. At the time, Houssay was the world's expert on hormones and researched in nearly every field of physiology, including blood circulation, respiration, digestion, the immune system, and snake and spider venoms, but his area of special interest was the hypophysis, also known as the pituitary gland.

The pituitary is a pea-sized gland that sits at the underside of the brain. It is protected in a small niche created by tiny skull bones called the *sella turcica* that form a shape resembling a miniature Turkish saddle. The pituitary secretes a number of hormones that control essential bodily functions such as how the body regulates blood sugar, provides pain relief, and regulates temperature, growth, blood pressure, water balance, saltiness of the blood, thyroid function, sex organ function, contractions during childbirth, and production of breast milk.

Houssay concentrated his research on the way the pituitary regulated carbohydrate metabolism. It had already been discovered that the pancreas, an organ behind the stomach, is the source of insulin. Insulin, in turn, drives the blood glucose into cells. It was also known that the liver responds to the body's need for energy by making or storing glucose. Houssay found that the pituitary affected blood sugar even more than the pancreas or liver by way of hormone secretions that greatly influenced the actions within those organs.

Houssay had an amazingly productive career, churning out more than five hundred scientific papers. He created a world-class research center in Argentina that attracted top international scientists. He was nominated for the Nobel Prize a total of forty-six times beginning in 1931 and finally won in 1947.

In 1943, a military coup in argentina resulted in a dictatorship, and Houssay was fired from his position as the head of the physiology department due to his prodemocratic views. He reestablished his research laboratory in a privately funded setting by founding the Experimental Medicine and Biology Institute, modeled after the Rockefeller Foundation and the Pasteur Institute. He fared no better in 1946 under the succeeding government of Juan Domingo Perón, who fired Houssay from the university a second time, probably for his antifascist views. Houssay was determined to stay in his home country and declined numerous offers to establish a research laboratory at foreign universities. He continued in his privately funded laboratory until the fall of Perón in 1955, when he was finally restored to his post at the university. He finished his career there and died at eighty-four years of age in 1971.

Carl Cori and Gerty Cori were both born in Prague in 1896, in what was then part of the Austro-Hungarian Empire (which later became Czechoslovakia). Gerty Radnitz was schooled at home until secondary school and then attended an all-girls lyceum. She decided on a career in medicine at the age of sixteen, but she lacked the necessary prerequisites for medical school. Over the ensuing year she completed study of the equivalent of five years each of math and science and the equivalent of eight years of Latin.

Carl Cori and Gerty Radnitz met at age eighteen when they both enrolled in medical school. They married upon graduation in 1920 and published their first joint paper the same year. Gerty worked on a pediatric ward for a couple of years, studying blood disorders as well as thyroid regulation of body temperature. Postwar food shortages were severe, and Gerty and Carl worked long hours in exchange for a single meal per day. She developed *xerophthalmia*, a severe dry-eye condition as a result of malnutri-

Gerty Radnitz Cori was the first woman recognized with the Nobel Prize in Medicine when she won with her husband Carl Cori for their work on how the human body metabolizes sugar.

tion, specifically, the lack of fat and vitamin A. Xerophthalmia can develop after just a few months of malnutrition and progress to cause ulcers on the eyeballs and blindness. It is still occurring in famine-stricken regions of the world today.

The Coris hoped their academic credentials would support their immigration to another country in which there were better economic and political conditions. Carl was recommended for a post at the State Institute for the Study of Malignant Diseases in Buffalo, New York, but after the interview, he thought nothing would come of it. Instead of waiting to find out, in 1921 he took a position in the research facilities at the University of Graz, Austria. The post–World War I fervor of racism affected all areas of life in Austria including the university, where a condition of Carl Cori's employment was to prove his Aryan descent.

Ironically, he was assigned to work in the laboratory of Otto Loewi in Graz. Loewi had recently conducted his famous experiments on dog hearts to prove the existence of neurotransmitters, for which he was awarded the Nobel Prize in 1936. But Loewi was born to a Jewish family, and in 1922, his future in Austria was questionable. Gerty also had Jewish heritage, although she had converted to Catholicism to marry Carl. The political environment made their move out of the country even more urgent. To his surprise, Carl was then offered the position in New York and granted a visa, but Gerty stayed behind for six months still waiting for her job offer. Despite her credentials (she had already published medical research papers), she was overlooked due to being a woman. She finally landed a low-level laboratory job at the institute and was able to follow her husband to America.

In New York, Carl began research on the mechanisms of energy production in the cells and focused on carbohydrate metabolism. Gerty conducted research on the effects of X-rays on the body. She managed to join Carl's research efforts in his lab after hours, for which they were both criticized as being an unacceptable distraction. At one point, the director of the institute threatened to fire Carl because of Gerty's involvement.

Gerty persisted and had the adamant backup of Carl. Their steady stream of important publications soon made it clear that they were collaborators on equal footing, each of their contributions essential to the forward progress of the research. Gerty was characterized as the more creative and aggressive laboratory expert with demandingly high standards, while Carl was more reflective and came up with the theories to test.

The Coris drew international recognition for their discoveries about metabolism of glucose in muscle and the liver. However, their work was not aligned with the cancer focus of the institute, which was the world's first comprehensive cancer research and patient care center and would later become known as the Roswell Park Cancer Institute. They sought research positions elsewhere, but Carl turned down offers from Cornell University and the University of Toronto because they refused to also offer Gerty a job.

The most attractive offer for Carl came from the Washington University School of Medicine in St. Louis, Missouri. Gerty was still subjected to sex discrimination. In fact, the university had to bend its own antinepotism rule to offer her a lowly research assistant job at one-tenth her husband's salary. In 1931, they moved to St. Louis. The department was not pleased with Gerty's participation in Carl's work. The Coris were warned that her involvement might be a detriment to his career. Again, the couple persisted, and Gerty often slept on a cot next door to the laboratory on the late nights she spent there.

Their prizewinning research consisted of the discovery of the exact biochemical changes that carbohydrates undergo to facilitate energy production in the body. They found that glucose in muscles was converted to lactate in a several-step reaction that liberated energy. The lactate left the muscle and traveled in the bloodstream to the liver, where it was taken up and recycled back into glucose. The recycled glucose traveled back to muscles when they needed it. This is known as the *Cori cycle*.

In 1943, Gerty was elevated to an associate professor. When Carl became chair of the Department of Biochemistry in 1946, he finally promoted her to full professor. Their laboratory attracted the brightest students and would eventually generate six subsequent Nobel winners, five in medicine and one in chemistry.

The Coris were on vacation in the mountains in 1947 when Gerty was diagnosed with a preleukemic condition of the bone marrow. *Myelosclerosis* (also called *myelofibrosis*) is a cancerous proliferation of a type of bone marrow cell that produces fibrous tissue. Scarring of the bone marrow crowds out normal new blood cell production. It is possible that during her brief early work, the effects of X-rays while at the research institute in New York might have resulted in excess radiation exposure, which is now known as a risk factor for this cancer. Two previous Nobel-winning women succumbed to similar diseases that were related to their X-ray exposures in research. Marie Curie (Nobel Prize in Physics, 1903, and Nobel Prize in Chemistry, 1911) died at age sixty-six of the related condition *aplastic anemia*: an inability to make red blood cells. Her daughter, Irene Joliot-Curie (Nobel Prize in Chemistry, 1935), was fifty-eight years old when she died of leukemia.

The Coris received notice of their Nobel Prize a few months after Gerty's diagnosis. There was no cure for myelosclerosis, but patients got blood transfusions as needed to improve well-being and prolong life. Carl drew her blood and tested it in his laboratory regularly, and he administered her blood transfusions. Gerty continued to work for ten years until her death at the age of sixty-one.

In 1960, Carl Cori remarried. He moved to Harvard as a visiting professor and continued research until one year before his death at the age of eighty-seven, in 1984.

37

The Making of a Silent Spring

Alfred Nobel's will stipulated that the prize would be awarded "to those who, during the preceding year, shall have conferred the greatest benefit on mankind." The most obvious way to measure benefit is lives saved or immediate deaths averted. It is easy to see how Paul H. Müller won the 1948 Nobel Prize for his discoveries relating to the insecticide DDT (*d*ichloro*d*iphenyl*t*richloroethane). In its initial widespread use, application of DDT immediately controlled deadly epidemics of typhus (spread by lice), malaria, and yellow fever (spread by mosquitoes) in several countries. Such fast and complete action by a man-made agent had never been seen before.

Paul Müller was born in 1899 in Switzerland and lived most of his life in Basel. His college education was interrupted when he went to work in industry for a couple of years, but he returned to school and earned his PhD in chemistry, graduating summa cum laude. He joined the staff of the Geigy dye factory, where his initial work was with natural and synthetic dyes and plant-based leather tanning agents.

In 1934, he was assigned to develop insecticides. Müller outlined the exact specifications of an ideal insecticide: it would work by killing on contact and have a quick and powerful effect on the largest possible number of insect species while causing little or no harm to plants and mammals. It would be cheap and long lasting with a high degree of chemical stability. Insecticides available at the time included pyrethrums and rotenone, relatively nontoxic insecticides derived from plants. Other insecticides were based on mercury, arsenic, and lead, and thus highly toxic.

Müller was strongly motivated by a recent severe food shortage in Switzerland, highlighting the need to protect crops. In addition, from 1918 to 1922, Russia experienced the most massive typhus outbreak in world history. The lice-borne illness began in the crowded cities and central provinces, then spread beyond the Ural Mountains into Siberia and central

Asia, engulfing the entire country by 1920. After the Russian Civil War ended, there was a brief lull before a famine struck and typhus returned with a vengeance, infecting countless numbers. Over a million were estimated to have died. Military forces in World War I often had more losses from malaria or typhus than from enemy fire.

Müller studied the pattern of chemicals that already showed some insect-killing activity and determined that effective substances contained a chloride molecule and a ring of six carbons (called a *hydrocarbon*). Indeed, Geigy's chief pharmaceutical chemist, Henri Martin, had previously made a chlorinated hydrocarbon insecticide that was effective against a type of moth. Müller systematically tested substances with variations of this basic formula. In 1939, when the 350th compound was applied to a glass testing chamber, the live flies placed into the box soon died. The sprayed chamber remained poisonous to flies for weeks. He had similar success with moths, beetles, and other kinds of flies. In 1940, it was tested by the Swiss government on crop-destroying beetles and found to be wildly successful.

DDT came to be characterized as a "slow knockdown" but "sure kill" insecticide. Müller had named chemical stability as a desired criterion of the optimum pesticide. The key to DDT's long life was that it was not water soluble. It was found to work on insects by first dissolving the fatty layer that normally repelled water from their surfaces. It then attacked the nerves, causing them to fire without control. The bugs fell on their backs and had convulsions until the DDT eventually paralyzed their nerve centers.

In a 1946 paper, Müller admitted that a very long-lasting poison would be disastrous for nature when he wrote:

> Pyrethrum and rotenone, like all natural insecticides, are completely destroyed in a short time by light and oxidation, as opposed to the synthetic contact insecticides which have been shown to be very stable. Nature must and will behave in this way, for what a catastrophe would result if the natural insecticide poisons were stable. Nature plans for life and not for death![1]

Müller was fully aware that a long-lasting poison would be disastrous for the environment.

Plant-derived pesticides like pyrethrums were short acting, waning in effectiveness within a few hours, and not effective at all after a day or two. DDT was much more stable, but the duration of its toxicity was not even guessed in the early days. It was eventually discovered that surfaces sprayed

1. T. N. Raju, "The Nobel Chronicles, 1948, Paul Hermann Müller (1899–1965)," *Lancet* 353, no. 9159 (April 3, 1999): 1196.

with DDT and stored under dust-free lab conditions were still toxic to insects seven years later. Experiments with dogs in 1944 and 1945 demonstrated that DDT concentrated in the fatty tissues of animals. More information on how long DDT persisted only gradually emerged.

Pyrethrums had been cheap and readily available until the outbreak of World War II, when the main supplier, Japan, halted shipping pyrethrum to the Allied countries. This added urgency to finding a synthetic insecticide. The landmark scientific paper describing Müller's synthesis of DDT and its effect on insects was published in 1944. It was soon realized that he was not the first to make DDT. German chemist Othmar Zeidler had synthesized the same compound in 1874 but had not found any useful application for it and left it on the shelf. This did not prevent Müller from patenting it on several continents.

Switzerland was politically neutral during the war and shared the patent with both sides. The US government tested it in 1943 and immediately contracted with Geigy and DuPont to manufacture it domestically. Müller also patented it in Britain and Australia and shared the formula with Germany. In addition to killing the vectors spreading typhus and malaria, DDT was eventually found to be effective in killing fleas, which transmitted bubonic plague; mosquitoes, which carried the parasite causing yellow fever, dengue, Chagas disease, elephantiasis, and viral encephalitis; tsetse flies, which carried the agent of African sleeping sickness; and sand flies, which carried the parasite of leishmaniasis.

The liberating Allied forces used DDT to eliminate a typhus epidemic in Naples, Italy, in 1944. Military police sprayed people who lined up in long queues, and spray teams in trucks dusted shelters and public buildings, ending the Naples epidemic in one month. The Allied liberators repeated their successful actions on concentration camp survivors, prisoners of war, and refugee populations, as well as in military barracks. Subsequently, it nearly eliminated malaria in Greece and in the Asia-Pacific war theaters. By the time of the 1948 presentation of the prize, the Nobel Committee said, "Without any doubt, the material has already preserved the life and health of hundreds of thousands." Since then, DDT has been hailed by some as the most valuable chemical ever synthesized to prevent disease.

Disputes arose regarding who should be given credit for DDT. The 1944 paper on DDT listed the lead author as Geigy research director Paul Läuger and the second author as chief pharmaceutical chemist Henri Martin. Müller's name came last. Although there was no argument that

DDT was synthesized and tested by Müller, Läuger asserted that he had directed Müller's work and pointed him in the right direction on which kind of compounds to test. The disagreement turned into a fight when the patents for DDT meant a lot of money for the rightful inventor, but Geigy's response to the squabble was to fire Läuger from the company in 1946.

Läuger was nominated for the Nobel Prize in 1946 and 1947. The Nobel Committee evaluated his worthiness for the discovery and decided that Müller was more deserving, but in those two years, no one had nominated him. They had to wait until 1948 when six Turkish physicians nominated Müller. He was the first prizewinner in medicine whose work was exclusively in a corporate setting rather than a hospital or academic environment and the first nonphysician to win the prize in medicine.

After the Nobel Prize, Müller complained to his employer that it was not paying him enough from patent profits. He retained an attorney and threatened to sue. Geigy responded by making Müller a vice director at Geigy and reaching a satisfactory agreement about the sharing of profits.

As early as 1945, the US Department of Agriculture (USDA) had selectively bred a strain of housefly in the laboratory that was relatively resistant to DDT. Increased resistance to DDT was fully predicted to occur in nature, and soon did. This provoked the use of even more massive doses of DDT and the creation of related spin-off chemicals that were hoped to be more effective. It was soon discovered that some insects developed a new protein that blocked the action of DDT. The same year, the USDA conducted an experiment in a forest in Pennsylvania: it sprayed an oil solution of DDT at a concentration of five pounds per acre to control a gypsy moth infestation. All moths died, and so did every bird (over four thousand), as well as ladybugs. Aphids, however, were not affected, and without their natural predators (the ladybugs), the aphids rapidly began to defoliate the forest. A fortuitous rain squall halted the destruction.

In another test, one-fifth the concentration of DDT (one gallon per acre) killed the moths and spared the birds but was still lethal to all aquatic life. DDT treatment of peach trees infested with moth caterpillars was found to be more lethal to a parasite that attacked moths than it was to the caterpillars. There were reports of treated fruit trees rapidly turning blood red with teeming masses of red spiders, whose natural enemies were wiped out by DDT. The USDA report described DDT as a "two-edged sword" that was promising and at the same time menacing.

Despite these known, documented hazards, DDT was released for public use later in 1945. That very year, the *Nebraska Farmer* magazine headlines

DDT spraying of children at public pool in San Antonio, Texas, in 1946

read "DDT—Our War Famed Bug-Killer," "The Wonder-Drug," and "This Joe Louis of the Insecticides."

It was sprayed on livestock, gardens, barns, and crops to get rid of grasshoppers, flies, moths, and mosquitoes. It was used on orchards and in forests. People applied it to the interior walls of their homes, which remained fatal to flies and mosquitoes three months after application. It was used in cellars, basements, yards, and residential schools to get rid of flies, fleas, roaches, ants, and bedbugs. A DDT-treated mattress would be fatal to bedbugs for nine months. Blankets could be washed a few times and even dry-cleaned and still kill any moth that came into contact with them.

DDT foggers were sprayed on children eating in school cafeterias, playing in public parks, and swimming in public pools. DDT was used extensively in American agriculture, especially on crops of cotton, peanuts, and soybeans. When Paul Müller visited the United States and was treated to a crop-dusting demonstration, he was appalled at the overuse of DDT and the lack of adherence to dosage instructions and warnings on the label. It is estimated that 1.5 million tons of DDT were used worldwide between the 1940s and 1970s.

DDT has been credited with eliminating malaria in the southern United States. In fact, demographic analysis showed that population movement away from malaria-prone regions was a more significant factor, as malaria rates had already been falling for more than ten years before DDT was introduced.

Rachel Carson was a marine biologist working as a nature writer for the US Fisheries Service. By the mid-1950s, she had already enjoyed financial success as a best-selling nature author. Carson had begun to track down information about the harmful effects of pesticides on wildlife, found the early USDA studies on DDT, and followed up on the subsequent scientific documentation of DDT's toxicities. Carson investigated in person many instances of DDT disasters and obtained the details of lawsuits filed in the 1950s by ornithologists and beekeepers about the toxicities of DDT on wildlife. She contacted scores of experts, including scientists from the National Institutes of Health (NIH) and leading cancer researchers.

Carson compiled the facts about DDT and presented them in her compelling writing style in *Silent Spring*, a three-hundred-page book published in 1962. Just as the gypsy moth eradication programs had also killed every bird in the forest, Carson warned of a day when there would be a silent spring due to no songbirds. She detailed the reports showing how DDT and its metabolites, DDE and DDI, accumulated in the fatty tissues of fish, birds, and all mammals from mice to polar bears. DDE and DDI were also found to be toxic to animals. *Silent Spring* detailed how DDT thinned the eggshells of birds and decreased survival of eggs and live-born chicks. She wrote about the findings of liver tumors in exposed animals and described the evidence for superbugs resistant to DDT and related derivative chemicals.

Silent Spring was credited with igniting the environmental movement in America. The book and interview of Carson were the subjects of a CBS-TV news special in 1963. The book led directly to Senate hearings in 1963 on the banning of DDT in the United States and to legislative control of the unregulated use of pesticides. The *New York Times* excerpted three sections of the book even before it was on bookshelves and featured headline announcements in the weeks before it hit the shops. It stayed on the best-seller list for thirty-one months.

Evidence supporting Carson's warnings continued to pour in. The year after *Silent Spring* was published, the cancer-causing effects were confirmed by Dr. Wilhelm Carl Hueper of the National Cancer Institute.

He reported DDT to be cancer-causing, and incriminated DDT in the production of benign and malignant tumors of the liver, cancers of the lung, and leukemias.

Biologists published reports of DDT toxicity year after year. For example, a study of red-tailed hawks and great horned owls in Montana in 1966 and 1967 described measurable DDT, its metabolite DDE, and newer, related pesticides DDD and dieldrin in every egg and tissue sample tested. Decreased chick survival was documented in those most heavily burdened with DDT. Subsequent studies have demonstrated that DDT and its degradation products acted as hormone interrupters resulting in the promotion of hormone-sensitive cancers. One study showed a higher risk of breast cancer in the children of women who were exposed to DDT when they were pregnant. In 1967, the Environmental Defense Fund was founded by scientists who began the litigation to ban DDT.

In 1968, it was estimated that one billion pounds of DDT and its toxic degradation products remained in the environment. It was banned for most uses in America in 1969, and in most other developed nations it has been at least banned from agricultural uses. In developing countries, DDT is still used to fight insect-borne diseases. Despite the ban, a 2010 study described computer simulation models that predicted substantial quantities of DDT were still being released from the world's oceans. Researchers at the Max Planck Institute for Chemistry in Germany reported that DDT was continually reentering the atmosphere from the ocean before being dissolved again in a recurring cycle. DDT is moving northward and expected to reach peak concentrations at the North Pole by 2030.

In the meantime, insect resistance to DDT has grown. In 2008, scientists discovered that resistant mosquitoes produced more of a specific protein that helped them metabolize DDT to a nontoxic form.

A safe alternative to toxic pesticides might be found in a lowly bacterium. *Bacillus thuringiensis israelensis* is a bacterial species that was discovered in Israel in 1976. It produced a substance that is toxic to mosquitoes and some kinds of caterpillars, gnats, and flies but is nontoxic to mammals, including humans. Unlike DDT, the bacteria did not persist in the environment.

Since 1960, Rachel Carson had suffered with metastatic breast cancer. She succumbed to complications of the disease in 1964 at age fifty-six, just two years after the book was published. Paul Müller died after a short illness the following year, at the age of sixty-six.

38

Scrambled Brains

Walter Rudolph Hess and Antônio Egas Moniz shared the Nobel Prize in 1949 for their experiments on the brain.

Walter Hess was born in 1881 in Switzerland and fell ill with tuberculosis, for which there was no treatment at the time, as a child. The family doctor who cared for him had a big influence on his decision to pursue a lifetime of medical study. Hess recovered and went on to complete medical school in 1906. Before he began to research the brain, he made several other scientific contributions. He developed a device to accurately measure the viscosity of blood, which could be an indicator of inflammation in the body. His testing device remained in use in hospitals worldwide for decades before a more compact, modern testing method was developed. However, Hess's original manuscript on the subject was rejected by a medical journal largely because he did not have a known scientific mentor to give his invention some prestige.

Hess advocated that patients with broken legs should be encouraged to get up and walk in their casts, as the stress of weight bearing promoted bone healing. This too was ignored since it went against the medical advice of the day, which was to rest for weeks. How soon a patient should resume activity after breaking a bone is still an area of conflicting advice. Some studies show that early weight bearing is fine after the bone is pinned in place with rods and screws, while fractures that involve a joint probably should not be engaged in weight bearing until after some bone regrowth has had a chance to stabilize the joint.

Hess next took up the study of ophthalmology, in which he invented a device to measure the deviation of a lazy eye. The Hess screen is still in use today. His thorough understanding of how the eyes work led him to develop a way to take three-dimensional photographs using prism lenses, but World War I interrupted his plans to patent it. He ran thriving ophthalmology practices in Zurich and two satellite offices and then left office practice to return to a study of anatomy, circulation, and respiration.

He temporarily filled the vacant position of head of the Physiology Department at the University of Zurich. When it came time to fill the position permanently, Hess's candidacy was initially passed over since he was not German. Germany was known as *the* prestigious country from which most major medical discoveries emerged, and European universities sought German staff members to enhance their status. This resulted in upset among Hess's students and trainees, and he ultimately gained the appointment. In one of his first actions as department head, he introduced the use of movies as training aids to study physiology in live animal subjects.

In the 1930s, Hess carried out a series of experiments on live cats that would later earn him the Nobel Prize. He devised tiny wires that were so small they could be accurately placed in specific positions in the brain. The area of the brain he studied is known as the *interbrain*, also called the *diencephalon*. He allowed the cats to walk about while a controlled microcurrent of electricity was run through the wires, and experimented with changes in the precise location of the wires and alterations in the strength, frequency, and duration of the electrical flows. He was able to provoke the cats to experience aggression, flight, fear, vigilance, stimulation, apathy, hunger, defecation, urination, sleep, and even coma.

The interbrain contains structures that are thought to be the first-ever brain parts formed by the evolutionary ancestors of animals on Earth: the thalamus and the hypothalamus. By using electrical stimuli, Hess was able to elicit basic, instinctual bodily functions along with some of the lower emotions that accompany them. Hess observed that electrical stimulation of the lateral regions of the hypothalamus caused apathy, sleepiness, and coma. Stimulation of the back portion of the hypothalamus led to excitement or could cause the animal to nervously try to run away or attack the laboratory attendants. Electric signals to the front part of the hypothalamus led to a general slowing down with a fall in blood pressure, slowed breathing, constriction of pupils, and slowed heart rate.

Hess was awarded the Nobel Prize "for his discovery of the functional organization of the interbrain as a coordinator of the activities of the internal organs." In later decades, these locations and functions in the cat brain were roughly correlated to the interbrain of humans.

Hess's experiments on live, unanesthetized animals provoked an outrage among antivivisectionists both locally and internationally. In fact, he was violating rules formed by a Zurich scientific committee that had negotiated with antivivisection advocates in 1895 to place limits on acceptable animal experimentation. These rules included injunctions against operating on unanesthetized animals and forbade repeated experiments on the same animal. Hess defended his practices publicly and did not cease his experiments.

The debate over animal experimentation continues to this day. Switzerland has some of the most stringent animal protection laws and in 1992 became the first country to include protections for animals in its constitution. Per Swiss regulations, the most important analysis is whether animals are needed in the first place: experimentation on animals is denied if the harm to animals is greater than the anticipated gain in knowledge. Second, the proposed animal experiments have to be based on sound experimental design and conduct. This requires the studies to be free of bias. The designs of studies using animals are evaluated by Swiss authorities to see if bias-reducing measures have been implemented, such as *blinding* (for example, the researchers not knowing if the animal is getting the active drug or a dummy pill), *randomization* (assigning animals at random to the treatment group and to the control group rather than selecting out certain animals for treatment), and *sample size calculation* (too small of a sample size makes the results questionable, while too large of a sample size unnecessarily uses excess animals).

A 2016 study looked at 1,277 approved applications for animal experiments in Switzerland and found that bias-reducing measures were rarely present, both in the applications describing the proposed experiments and in the resulting scientific publications. The study authors concluded that the Swiss authorities were not following their own rules to protect against unnecessary animal experimentation.

Hess was working in an era when a tremendous number of new discoveries about the brain were being made and many researchers were seeking to find the anatomical correlates of personality, consciousness, and thinking. He emphasized that science should always acknowledge the limits of scientific discovery. In fact, he refused to rule out the possibility of unknown powers and effects, which is, after all, good scientific method. He had unconditional respect for religion, although he was not religious.

Antônio Egas Moniz

In contrast, the lack of physical correlates for anything but instinctual behaviors did not cause the slightest hesitation for Hess's Nobel cowinner, Antônio Egas Moniz, who practiced psychosurgery in an attempt to eliminate unpleasant emotions, thoughts, and behaviors by random destruction of brain tissue. Antônio Caetano de Abreu Freire Egas Moniz was born to an aristocratic family in Portugal in 1874, attended medical school in France, and returned to Portugal in the Department of Neurology at the University of Lisbon. His first fame came from his 1901 publication of a series of essays called *A vida sexual* (The sexual life). Among other subjects, Moniz took the position that the race could be improved by negative eugenics, which prevented breeding among those with "morbid heredity": undesirable inherited characteristics.

He advocated birth control or sterilization for those women considered mentally ill, poor, and undereducated; whose sexuality was considered irresponsible; or who represented bad moral behavior (such as prostitutes). In contrast, healthy middle-class women were advised to procreate freely. Moniz described homosexuality as resulting from an inherited predisposition that made a person susceptible to certain situations in life. An example Moniz gave of an environment promoting homosexuality would be attending an all-boys boarding school. He described that such genetically predisposed individuals would not be homosexuals if they were not dragged down into a "degrading vice" by the environment. He regarded homosexuality as a disease and advised psychiatric treatment for it, for the survival of the individual and of society.

A vida sexual was officially available only by prescription but became an instant best-seller on the black market, achieving its tenth edition in 1933. Moniz wrote a biography of a priest who was also a physician and a hypnotist and ultimately became a pope. Oddly, Moniz wrote a history of playing cards too.

For the next couple of decades, Moniz focused on politics and was briefly jailed on three occasions in political power clashes. He served more than ten years in the Portuguese parliament and acted as Portugal's ambassador to Spain during World War I. Although Spain remained neutral in World War I, Portugal sided with the Allies to protect its African colonies from the Germans. Moniz represented Portugal by signing the Treaty of Versailles for his nation, which allowed it to retain the African properties and also gave it a few pieces of German-held borderlands in Africa. He resigned from politics after a military coup in 1926.

Long before his attempts at psychosurgery, Moniz earned fame for inventing a method to visualize the blood vessels of the brain in a living patient. Researchers in Vienna had some slight success by injecting cadaver heads with a mixture of lime, mercuric sulfide, and petroleum to cause vessels to show up on a head X-ray. Moniz sought a less-toxic approach that could be used in live humans and started by studying the properties of bromides. The element bromide is a brown liquid at room temperature and known to be toxic and corrosive when it is not diluted. It has properties between that of iodine and chlorine. Diluted bromide-based drugs were used as sedatives in the nineteenth and early twentieth centuries, so Moniz figured that they accumulated in the brain and, in sufficient concentration, would show up on X-ray film.

Initially, he studied cadaver heads, but his medical center could not supply enough for his research needs. He contracted with a morgue across town to sever the heads of corpses and transport them to his laboratory in the trunk of his limousine. After the experiments, the heads were carted back to the morgue. As his work progressed to live human subjects, he tried giving bromides by vein, but the X-rays remained black. Finally, he gave bromide by direct injection into the carotid artery in the neck in a series of six patients. In the sixth patient, the cerebral blood vessels were faintly visualized in an X-ray of the head, but the patient developed a blood clot as a result of the injection and died.

Moniz switched to using iodine, which had a better chance of showing up on X-rays. In 1927, he had his first success with iodine injection, resulting in the outline of brain blood vessels in X-rays. In subsequent years, he wrote over a hundred papers on the procedure and its refinements. Today, cerebral angiography is still performed using an iodine compound; other elements that show up in X-rays are used in studying the gastrointestinal tract (barium) or when doing an MRI (gadolinium). Moniz expected to win the Nobel Prize for his development of cerebral angiography, and he was nominated eighteen times starting in 1928. He ultimately won for an entirely different procedure.

Moniz next turned his attention to the treatment of the insane. In November of 1935, he and his assistants made the first attempts at psychosurgery. They treated seven unlucky mental patients by drilling holes in their skulls and injecting absolute alcohol. There is no distinct record of the fate of these patients as Moniz's record keeping was very poor. He wanted more definite disruption of nerve connections, so he eventually devised a needle-like instrument with a retractable wire loop, which he called a *leucotome*. He would insert it through the drilled skull holes and vigorously pass this instrument back and forth, slashing through brain tissue. He named the procedure *leucotomy* (from the Greek *leu*, referring to the white matter of the brain, + *tome*, "cut," "slice").

In 1936, Moniz published results of his first twenty leucotomy operations on patients in which he claimed that within a few days, the patients were "improved." His first patient was less agitated and less paranoid, but she was also nauseated, disoriented, and more apathetic, and she became dull. Moniz did not do any IQ tests or memory interviews and did not provide any evidence for his assertion that memory and intelligence were unaffected by his operations. He put patients who suffered severe negative effects in the "unchanged" category, but even called some of those "improved" because in their zombie-like apathy, they were indeed less agitated than before the operation. As far as can be re-created from surviving medical records, many of the patients became catatonic.

Moniz would later come under attack for inadequate documentation, poor patient follow-up, and grossly understating adverse effects. He admitted in a 1937 interview reported in the *New York Times* that it was possible to slice away too much imagination. Impartial observers documented horrific personality loss in leucotomy victims.

In 1948, a Swedish professor of forensic psychiatry reported the mother of a leucotomized child as saying, "She is my daughter but yet a different person. She is with me in body, but her soul is in some way lost."[1] A psychiatrist who submitted some of his patients as Moniz's guinea pigs later called Moniz's claim of improvement "pure cerebral mythology." According to a 1949 medical paper, postsurgical patients were routinely described as dull, apathetic, listless, without drive or initiative, flat, and lethargic. They were

1. Bengt Jansson, "Egas Moniz", The Nobel Prize, October 29, 1993, https://www.nobelprize.org/prizes/medicine/1949/moniz/article/.

placid and unconcerned, docile, passive, and lacking in spontaneity. These patients were without aim or purpose, childlike, and dependent.

American neurologists Walter Freeman and James Watts at George Washington University Hospital eagerly obtained the initial paper by Moniz. Freeman headed the Neurology Department, and Watts was chief of neurosurgery. In 1936, they ordered a leucotome and went right to work on live human subjects picked up from mental hospitals and taverns. They renamed it *lobotomy* in recognition that it was causing broad destruction of the frontal lobes. Freeman thought that mental illness was caused by unwanted emotions and that he could cut away the physical seat of those emotions. He explained that this operation separated the emotional brain from the thinking brain.

The immediate death rate from lobotomy was 14 percent; however, the eventual death rate from complications of surgery was never accurately recorded, not to mention the toll of ruined lives. After operating on eighty patients, Freeman and Watts published a book on psychosurgery in 1942. They admitted, "Every patient probably loses something by this operation, some spontaneity, some sparkle, some flavor of the personality."[2] That year, Freeman wrote to the head of the Veterans Administration (VA) hospital system requesting that lobotomy be offered freely to all veterans with mental illness. It was immediately approved, and eventually, over two thousand veterans were lobotomized.

Freeman and Watts soon decided that the flimsy leucotome wire was not affecting enough tissue. They developed a new approach to the frontal lobes by inserting an instrument under the ridge of the eyebrows. Various implements were tried, including an apple corer, a butter spreader, and an ice pick. These were rotated and swept back and forth to gouge into the maximum amount of tissue. By 1950, Freeman and Watts extended the indications for lobotomy beyond mental patients to children with bad behavior, patients with chronic pain, and people with social anxieties. They had personally lobotomized over six hundred patients and published an expanded edition of their book.

Lobotomy became very popular among psychiatrists, who were encouraged to perform lobotomies even though they were not trained in surgery. "It is so simple," said Freeman, "and only requires local anesthetic or none

2. Walter Freeman, James W. Watts, and Thelma Hunt, *Psychosurgery: Intelligence, Emotion and Social Behavior Following Prefrontal Lobotomy for Mental Disorders* (Springfield, IL: Charles C. Thomas, 1942).

at all." Freeman himself usually operated without gloves or a mask. Some one hundred lobotomies were performed in the United States in 1946, but the number swelled to five thousand by 1949, the year that Moniz won the Nobel Prize for developing leucotomy psychosurgery.

Lobotomy became acceptable, even recommended for use on the socially undesirable. A 1946 *New York Times* headline read: "Operate on Brain to Reform Woman; Detroit Doctors Report Unusual Surgery Which May End Her Criminal Tendencies." The article described lobotomy as a "delicate brain operation, performed in the hope of turning a morally degenerate woman into a useful member of society."[3]

Lobotomies were used by the US prison system as a disciplinary measure.

The horrific effects of lobotomy were dramatically characterized by Jack Nicholson in the 1975 movie *One Flew Over the Cuckoo's Nest*, based on a 1962 book of the same title by Ken Kesey. Kesey had worked the graveyard shift at a Veterans Administration mental hospital in Menlo Park, California, during the height of the brief lobotomy era. The movie won five Academy Awards: Best Picture, Best Actor, Best Actress, Best Director, and Best Adapted Screenplay.

The most forthright description of lobotomized persons can be found in a heavily redacted CIA report on the use of the procedure on political enemies of the Communist state in Czechoslovakia: "This relatively simple operation on a sane person rendered them incapable of exercising their own will and made them mere robots." Also per CIA records, therapeutic lobotomy was banned in Poland by the USSR "because it was a mutilating operation with very poor results."[4]

Probably the most famous political victim of lobotomy was Eva Perón, the fiery and outspoken wife of Argentine dictator Juan Perón. She had chronic pain from metastatic cervical cancer, but the real reason for her secret lobotomy was likely the political moves she was making without consulting her husband, firing up a revolution into a potentially bloody civil war. The procedure was done under strict security by renowned Boston neurosurgeon Dr. James Poppen.

Another notable victim of lobotomy was Rosemary Kennedy, eldest daughter of Rose and Joseph Kennedy and sister of future president JFK.

3. Walter W. Ruch, "Operate on Brain to Reform Woman; Detroit Doctors Report Unusual Surgery Which May End Her Criminal Tendencies," *New York Times*, December 7, 1946.

4. CIA Information Report, date removed, sanitized copy approved for release in 2011, no. RDP80-00809A000600030562-5, https://www.cia.gov/library/readingroom/document/cia-rdp80-00809a000600030562-5.

She was mildly intellectually impaired, so slightly that it was not noticeable in usual social situations, not even by her early tutors. Her deficits became more noticeable as she grew up. Rosemary failed to progress in school beyond about fourth grade, but what aggravated her parents were her teenage outrages described as uncontrolled aggressions. Her behaviors presented the constant threat of exposure that the Kennedys had an abnormal child.

When Rosemary was twenty-three, her father took her to the best: Freeman and Watts at GWU. The surgery was a disaster. Rosemary went from slightly intellectually challenged to totally incapacitated. She needed extensive physical therapy for months to regain partial ability to walk and talk. She was shuttled to a long-term-care facility staffed by Catholic nuns in Wisconsin, where she lived in obscurity for six decades until her death in 2005.

Freeman became a maniac advocate, traveling the country in a "lobotomobile" and offering same-day procedures in the cities and towns he visited. He eventually performed lobotomies on upwards of twenty-five hundred patients in twenty-three states and on patients as young as four years old. A typical outcome was experienced by Anna Ruth Channels, who underwent a lobotomy for severe headaches in 1950. She never again complained of headaches, but she was left with the mind of a child. According to her daughter, Carol Noelle: "Just as Freeman promised, she didn't worry." Freeman's career screeched to a halt in 1967 when he operated on a housewife named Helen Mortenson who died of a brain hemorrhage.

In 1939, Egas Moniz was shot by a former patient and suffered a spinal injury, causing him to remain in a wheelchair for the rest of his life. He did not attend the Nobel ceremonies in his honor. He died in 1955 at the age of eighty-one. There is an unusual article on the Nobel Prize website defending psychosurgery and the choice of Moniz for the prize. It describes leucotomy as a delicate, exact surgical technique and exalts its benefits compared to other mental treatments of the era.

39

More Hormones

The 1950 Nobel Prize in Medicine was shared among two chemists and a medical doctor. Edward Calvin Kendall, Tadeus Reichstein, and Philip Showalter Hench were awarded the prize for their discoveries about adrenal hormones.

Kendall was born in Connecticut in 1886 and received his bachelor's, master's, and PhD degrees in chemistry from Columbia University. His first job in 1910 was with the Parke-Davis and Company pharmaceutical firm. Kendall was to isolate the thyroid gland hormone. It had already been discovered that patients who had had their entire thyroid glands surgically removed were at risk of cretinism from lack of thyroid hormone (see details in chapter 9). Patients without a thyroid gland could be given the ground-up thyroid gland of pigs to prevent cretinism. Kendall sought to find the pure hormone so that it could be pharmaceutically manufactured.

Kendall had not completed his work when he quit Parke-Davis in frustration over the intellectual isolation of working in a laboratory at a drug manufacturer. He applied for a position at the Rockefeller Institute and was bluntly turned down. He continued his research on the thyroid hormone at a new laboratory at St. Luke's Hospital in New York City, initially without any salary. His seniors there did not appreciate the importance of his thyroid work and instead tried to assign him the task of conducting a chemical analysis of breakfast cereal. At that point, Kendall promptly quit his job and soon accepted a research position at the Mayo Foundation in Minnesota.

Kendall's isolation of pure thyroid hormone (*thyroxine*) was the result of an accident: he was using alcohol to digest some hog thyroid gland and fell asleep at the lab bench. When he woke, the process had gone on too long and all of the ethanol had evaporated, leaving only a white, crusty material surrounded by a yellow waxy substance in the bottom of the beaker. He worked on this white substance for two more days and was successful in isolating pure crystals of thyroxine on Christmas Day 1914.

It took over ten years and the considerable work of many other chemists to sort out the exact molecular structure of the hormone, accomplished in 1926. Kendall's isolation of thyroxine made it possible to mass-produce a pure thyroid hormone supplement. It treated not only postsurgical patients without a thyroid gland but also patients who naturally developed under-active thyroids (*hypothyroidism*). Today in the United States, between 4 percent and 5 percent of the population has a low thyroid count. About 2 percent is taking some form of thyroid hormone. We now know thyroxine as *T4* for its four iodine molecules. The body converts T4 to triiodothyronine (*T3*, containing three iodines). T3 is the active form of the hormone that cells can use.

Thyroid hormone is available as a synthetic preparation called *levothyroxine*, which has become the second-most-used prescription in the United States with over 20 million filled prescriptions each year. However, up to 40 percent of patients still have hypothyroid symptoms while on levothyroxine. This is because they do not efficiently convert T4 to T3. Such patients do better on the old-fashioned natural extract from a pig's thyroid gland, consisting of a combination of T4 plus T3 in addition to small amounts of T2 and T1.

At Mayo, Kendall rapidly worked his way up to the directorship of the Biochemistry Department. Subsequently, he worked on a team that sorted out the structure of *glutathione*, the major biochemical mammals use to prevent cell damage from foreign substances and from the body's own toxic by-products of everyday metabolism. Glutathione has more recently been called the "mother of all antioxidants" and "the great protector." Kendall and his team found that glutathione is a simple compound consisting of three amino acids: L-cysteine, L-glutamic acid, and glycine. Glutathione is critical in preventing cancer and supporting a well-functioning immune system. Elevated levels of glutathione promote disease resistance and a long life span. Taking glutathione supplements by mouth is useless since they get broken down into their constituent amino acids in the gut. The liposomal form of glutathione is suspended in tiny fat globules in an effort to evade the digestive process and go right into the bloodstream. However, the cellular glutathione levels don't change much with this preparation either. Some foods can supply the enzymes and amino acid precursors that help the body to make glutathione. These include broccoli, cabbage, cauliflower, brussels sprouts, beets, raw milk, tart cherries, and the spices turmeric, cinnamon, and cardamom. It is difficult to consume these foods in sufficient quantity to significantly raise glutathione levels. An alternative to stuffing oneself with whole foods is taking the combination of un-denatured whey proteins, melatonin in its natural state in tart cherry concentrate, alpha lipoic acid, milk thistle, and MSM (methylsulfonylmethane). To optimize the body's production of glutathione, supplementation is given with vitamins C, E, B_1, B_2, B_6, folate (B_9), B_{12}, and the minerals selenium, magnesium, and zinc.

Kendall next turned his attention to isolating the hormone in the *adrenals* (from *ad*, "on top," + *renal*, "the kidney"). The adrenals are like small, sloping, triangular sacs capping each kidney. The glands are two-part structures consisting of an outer ring of tissue enclosing an inner portion. The two areas have totally independent functions. The outer ring (*cortex*) contains hormones that sustain life. The inner portion (*medulla*) contains hormones that maintain proper salt and water balance. The focus of Kendall's research was isolating the active substances in the adrenal cortex. This was tedious biochemistry work that required an entire team at Mayo. It was also being done under the immense pressure of funding being canceled if there were no results and of international competition from teams at Columbia University and the University of Zurich. Each group was discovering that many hormones were in the adrenal cortex, not just one. They independently named their discovered substances with letters of the alphabet, but one group's substance A was another's substance B, and so on. At Mayo, Kendall had two extracts that looked promising: compounds A and E.

World War II suddenly brought the keen interest of the US government to Kendall's research. There was a rumor that the Germans were buying up vast quantities of cow adrenal glands from South America to give adrenal extract to its Luftwaffe pilots. The implication was that adrenal extract would make the pilots resistant to the physiologic stressors of long flight missions and low oxygen. Indeed, it was known that animals and people with missing or damaged adrenal glands had a very poor tolerance to stress of any kind: physical trauma, infections, surgery, or even emotional stress could kill them. So, it was logical to think that supplementing healthy, stressed persons with adrenal extract might make them perform better. The American government's Office of Scientific Research and Development gave top priority to isolation of compound A in Kendall's laboratory, and the Mayo team worked in conjunction with researchers at the Merck pharmaceutical company to produce the substance. Eventually, the information about German adrenal supplementation was shown to be rumors, and after four years, the Mayo team was not seeing good results in patients dosed with its adrenal extracts, so the project was all but abandoned. Those years were also hard on Kendall personally. His wife suffered from emotional instability, one son who was in medical school died from cancer, and another son committed suicide after discharge from the military. Kendall continued his work by leading a much smaller team, now focusing on compound E, which he correctly guessed was the main life maintenance hormone made by the adrenals. By 1948, compound E was made in quantity by Merck. Kendall suggested that compound E could be tested to treat such diverse conditions as cancer and mental illness.

Meanwhile, also at Mayo, Dr. Philip Hench was working in the Department of Rheumatology. Hench was born in 1896 in Pennsylvania, where he completed all of his schooling, including medical school. From there he joined the Mayo Foundation, where he spent his entire career, except for a brief stint at the University of Freiburg in Germany and service at a military hospital in Arkansas during World War II. He was desperately seeking a way to treat the inflamed joints of rheumatoid arthritis (RA) sufferers. Hench observed that persons with RA have temporary improvement in their symptoms when they become pregnant, get yellow jaundice, or undergo surgery of any kind. Hench and other leading rheumatologists followed the work of Hans Selye, a Hungarian-Canadian researcher who wrote extensively on the role of the adrenal glands in assisting the body to deal with stress. Hench realized these major body stressors were probably causing a surge of adrenal hormones, with the side benefit of helping ease the symptoms of RA.

Dr. Hench requested Kendall's compound E, for which they coined the name *cortisone*, to test on patients. A large dose of cortisone was given to a female patient with remarkable improvement in movement and diminished pain. It was then tested on fourteen more RA patients, all of whom showed improvement. It soon became evident that symptoms were relieved only so long as cortisone was being given. Patients suffered a relapse as soon as it was withdrawn. Hench realized from the very beginning that cortisone was not a cure but provided some control of symptoms. He saw that cortisone suppressed the reaction of the joint tissue to whatever was irritating it but did not have any influence on the cause of the disease.

Some 4,465 miles away in Basel, Switzerland, Tadeus Reichstein was also trying to find the key adrenal hormone. Reichstein was born in Poland in 1897 and later immigrated with his family to the Ukraine and then Germany, ultimately settling in Switzerland. He became fluent in four languages and focused his studies on chemistry. His early work concerned the chemical compounds that gave flavor and aroma to coffee and chicory. Next, he developed a process to artificially synthesize vitamin C at the same time that the English chemist Norman Haworth made the discovery. Haworth received the Nobel Prize in Chemistry in 1937 for his work while no one even nominated Reichstein. Yet the industrial-production procedure for vitamin C, which bears the name Reichstein, is still in use in vitamin manufacturing today.

Reichstein then immersed himself in the study of adrenal hormones and played scientific leapfrog with the Mayo team, systematically isolating one of twenty-seven adrenal hormones after the other and many times in advance of Kendall's group, sometimes just behind them. He isolated cortisone at exactly the same time as Kendall. In his Nobel acceptance speech, Reichstein admitted that the clinical application of cortisone to arthritic disease was the Americans' idea, while he was only working in pure chemistry without any firm notion of how it could be used in patients. Eventually, cortisone and its derivatives would be used for treatment of allergic disorders, skin conditions, ulcerative colitis, arthritis, lupus, psoriasis, and asthma. It is also given to quell nausea in patients receiving chemotherapy.

After isolating all the adrenal hormones, Reichstein shifted his focus to seeking out other natural sources of cortisone than beef adrenal glands. In 1947, he directed an extended trip to Africa to collect the seeds of various medicinal plants that contain a cortisone-like chemical called *sarmentogenin*. By the time of his 1950 Nobel acceptance speech, he was still growing and testing the seeds of scores of plants. Eventually, sarmentogenin derived from seeds became one of the raw materials in the synthetic production of cortisone.

Reichstein continued chemical research for a few years after his official retirement from the university at the age of seventy. When he was seventy-five, he announced his plan to devote the next phase of his life to the study of ferns. He wrote over forty scientific papers on ferns. In a 1985 interview, eighty-eight-year-old Reichstein was asked for his secret to longevity. He replied, "There's no secret, but I think that as long as you are interested in life, and you can work, this is probably the best you can do for yourself."[1] In 1992, Reichstein and sixty-three other Nobel laureates signed an appeal to the world's governments to end fighting in Bosnia and Herzegovina. He lived to age ninety-nine.

Today, RA flares are still treated with cortisone or related steroid hormones. Every effort is made to limit their chronic use because of the rapid development of dangerous side effects. Patients on chronic corticosteroids develop central obesity, a fatty puffiness of the face called *moon facies*, muscle weakness, excess facial and body hair, high blood pressure, diabetes, cataracts and glaucoma, osteoporosis, neuropsychiatric problems, lack of menstrual periods, purple stretch marks, and easy bruising.

1. S. Sterkowicz, *Tadeus Reichstein—Life and Scientific Work* (Wloclawek, Poland: Wloclawek Scientific Society Press, 1995).

40

Nobel Influences

There were a number of strong influences on many of the Nobel Prize–winning researchers and their discoveries in the first half century of the contest.

The early years focused on infectious diseases, as they were common and death rates were high in those days. Nobel laureates themselves sometimes suffered, including the first prizewinner, Emil von Behring, who had tuberculosis. Charles Laveran and Robert Ross both came down with malaria while studying it. Niels Finsen died from heart failure probably due to a tapeworm infection. Ramón y Cajal had both TB and malaria. Robert Koch's second wife caught malaria on their world travels to hot spots of infectious disease. Ilya Mechnikov lost his first wife to TB and his second wife to typhoid, caught TB and recovered, and had an episode of self-inflicted relapsing fever. Mechnikov's cowinner, Paul Ehrlich, also had TB, requiring two years of convalescence in Egypt. Charles Nicolle won the prize for his discoveries regarding lice-transmitted typhus but died from infection with his later research interest, mouse-transmitted typhus. Christiaan Eijkman caught malaria on his first military deployment to Dutch Indonesia and upon his recovery was assigned to return there to study beriberi. Walter Hess had TB as a child and his attending doctor had a great influence on his subsequent career path. Hans Spemann also had TB and, during his long recovery, he read a book by an evolutionary biologist, effectively determining his future in embryology research and prizewinning discoveries that eventually led to cloning.

War and political events significantly affected many of the winners, and the Nobel Prize itself was a product of war; the money was initially generated from munitions manufacture and inventions of more lethal warfare. So did war often affect scientific research and the prizewinners.

Emil von Behring was a medic in the Prussian Army, which drove him to study diphtheria that killed more soldiers than enemy fire.

Robert Ross sought a post as a British medical officer in the Indian colonial force specifically to study malaria.

Ramón y Cajal would likely have made his discoveries much earlier if he had not reluctantly enrolled in the Spanish Army, serving as a medical officer in a Cuban regiment. He nearly died of malaria on his slow ship back home to Spain where he saw many a body slipped overboard in sailors' funerals. Ilya Mechnikov fled Russia due to lack of academic freedoms under unstable czarist rule and permanently relocated to France.

Ivan Pavlov applied his knowledge of physiology to assist the Soviet State in mind control of its citizenry.

Hermann Muller was a Communist-influenced academic when he helped in the distribution of an illegal newspaper on the campus of the University of Texas. He next went to Germany, where his oratory was ransacked by the Nazis. From there he set up genetic research laboratories in Russia where he had hoped to experience the Communist ideal but soon found the political environment too suppressive. He enlisted on the Communist side of the Spanish Civil War as a politically acceptable means to escape Russia. When the revolution fizzled, Muller managed to slip out to Edinburgh and then make his way back to America. There his political activism shifted to antinuclear efforts, and he remained a person of interest to the FBI.

Charles Laveran was a third-generation military physician faced with a greater loss of troops from malaria than from battle wounds. He directed his attention to research on the cause of malaria.

Róbert Bárány had already concluded his experiments on the balance mechanisms of the inner ear when he was assigned as a medical officer in the Austrian Army and sent to the Russian front. There he developed life-saving methods of treating head wounds. He had become a Russian prisoner of war by the time he was notified of his prize for ear research, and it required the intervention from the king of Sweden and the International Red Cross to free him.

Julius Wagner-Jauregg was a military doctor in World War I and involved himself in the torture of deserters and electroshock of battle-fatigued veterans. His Nobel Prize–winning treatments of mental patients with malarial blood was only a shade kinder.

Charles Richet had served as a medical corpsman in the Franco-Prussian War in 1870 before his research led to a prize for discovering the mechanism of allergic reactions. He later lent his prestigious endorsement to the field of eugenics, eventually becoming the president of the French Eugenics Society. Richet died in 1935, but in World War II, five relatives, including his son, were the victims of racial cleansing: they were put in concentration camps or other German prisons.

Paul Ehrlich sustained anti-Semitic attacks for developing a treatment for syphilis.

Otto Fritz Meyerhof and his family had to dodge Nazi authorities before being assisted to immigrate by the American-based Emergency Rescue Committee.

Otto Warburg volunteered for the Germany Army in World War I, in which he was injured and earned the Iron Cross, buying him a ticket out of the army so that he could commence his research on how cells get energy. Although he had Jewish heritage, Warburg was allowed to stay on at his research lab in Germany through the Nazi reign. Some think this may have been arranged by Hitler, who wanted to support Warburg's cancer research out of terror of the disease. At least three Warburg extended family members died in concentration camps.

A. V. Hill was an outspoken critic of Nazi concentration camps, even in the prewar years, and assisted in the relocation of many persecuted scientists and academics fleeing from Germany. Hill was a British military officer in World War II. His status as a Nobel winner gave him considerable influence when he served as ambassador on a secret project to persuade the Americans to enter World War II.

Karl Landsteiner had already completed his prizewinning research on blood types when rising anti-Semitism prompted him to leave Austria in 1919. He eventually settled in America but is said to have lived the rest of his life in fear of a Nazi takeover.

George Minot served as a surgeon for the United States Army in World War I, which promoted his interest in blood disorders. His cowinner, William Murphy, dropped out of medical school for lack of funding and enlisted in the United States Army for the last two years of World War I before obtaining a scholarship to continue his studies at Harvard.

Albert Szent-Györgyi served as a medic in the Austro-Hungarian Army (allied with Germany) and shot himself in the arm to qualify for medical leave after three miserable years in the trenches. After the Nazis occupied

Hungary, he assisted the Resistance by helping many Jews secretly leave the country. He was eventually hunted by the Gestapo and spent the last two years of the war in hiding. He donated the gold from his Nobel medal to Finland to help finance its resistance to the occupying Soviets.

As a teenager, Gerhard Domagk served in the German Army in World War I, in which he was wounded and transferred to the medic division. He won the Nobel for creating sulfa antibiotics at a German chemical company but was forbidden by the Nazis from accepting his prize.

Alexis Carrel's lifelong interest in perfecting surgical repair of blood vessels was supposed to have come from his distress at how the French president bled to death two days after being stabbed by an assassin. Carrel was another eugenicist and labeled as a collaborator with the Germans but died before the war crimes trials.

Gerty Cori suffered malnutrition in post–World War I Prague so severe that she developed xerophthalmia that threatened blindness.

Paul Müller was deeply affected by the post–World War I crop failure from potato blight that prompted famine in Switzerland, fueling his search for an effective pesticide. DDT was an instant success. Armies were the number one customers for DDT, and government supply contracts provided much-needed funding for the Nobel scientists to work alongside pharmaceutical companies in the development of effective large-scale production techniques. Similarly, the research on large-scale penicillin production was driven by massive orders from the United States Army.

Edward Calvin Kendall's prizewinning cortisone research was endorsed by the US government out of concern about rumors that the Germans were doping their pilots with performance-enhancing hormones.

Some prizewinners were well ahead of their times. Niels Finsen won the third Nobel Prize for successful light therapy, a treatment that was outlawed in the United States within a few decades. Recently, there has been a scientific revival in the study of therapeutic light. One of the 1908 winners, Ilya Mechnikov, promoted the idea of eating probiotic foods such as yogurt to balance gut bacteria, although it did not catch on for another seventy-five years. Otto Warburg won for his work on how cells make energy, but his related cancer theories proved too novel for consideration by the Nobel Committee. He hypothesized that a defect in cell respiration was the primary cause of cancer, and his theory was strongly debated at the time. Today there is renewed interest in studying the so-called Warburg Effect in tumors. Hermann Muller was awarded the 1946 prize for showing that X-rays can cause dangerous genetic mutations. He advocated for extremely conservative use of X-rays in medicine, but his advice has only very recently been heeded. Some of today's specialty societies have belatedly recommended that patients and doctors keep track of cumulative medical radiation dosing to ensure that exposure does not exceed known safe limits for a lifetime.

The Nobel Committee does not make any claims to fairness, but it does advocate a sense of righteousness and has a history of sticking to its decisions despite the existence or later emergence of conflicting evidence. In the first fifty years of the prize, there were notable flops. Julius Wagner-Jauregg's theory of using the blood of malaria patients to drive syphilis from the brains of demented people had insufficient proof when it was first reported. His very small study with poorly documented results was not adequately scrutinized by the Nobel Committee. Before long it became obvious that malaria did not treat syphilis at all but could kill the patient. Similarly, Johannes Fibiger's prizewinning theory that stomach cancer was caused by roundworms was not adequately studied by the time he was awarded the prize and soon proved false. Egas Moniz's invention of cerebral angiography was a remarkable contribution. His other bright idea, lobotomy, earned him the Nobel Prize, although it was responsible for one of the grimmest chapters in the history of Western medicine. Within just a few years, respectable physicians worldwide distanced themselves from any endorsement of the barbaric procedure.

Some Nobel-winning researchers made serious scientific blunders. The very first winner, Emil von Behring, was recognized for his research on diphtheria but also promoted the erroneous idea that human TB came from milk and advocated using formaldehyde to treat milk. Luckily, this poisonous solution was rejected because the odor was unacceptable to consumers. Robert Koch's famously ineffective TB treatment (tuberculin) was prematurely announced to bolster German scientific superiority, but he had accomplished so much in other areas of infectious disease research that the tuberculin embarrassment was overlooked.

Camillo Golgi won for his method of staining the nervous system so it could be accurately studied and then failed to study it. He proceeded to promote the erroneous idea that all nerves were connected in one jumbled network. Golgi embarrassed his cowinner, Ramón y Cajal, by giving a Nobel lecture on this erroneous theory when it had already been long since disproved by more careful observers like Ramón y Cajal, who used the Golgi stain to study the brain.

Emil Theodor Kocher won for thyroid surgery, and his meticulous surgical techniques are the foundation of safe operations. But he discovered to his horror that his total thyroidectomy operations were routinely making cretins out of his patients because the (then-unknown) thyroid hormone was lacking. Kocher worked closely with many others to quickly find solutions, which included leaving just a little bit of the thyroid gland behind or implanting some animal thyroid to substitute for the lost natural gland.

Paul Müller was immediately recognized for the amazing effectiveness of DDT in wiping out typhus at the end of World War II. In fact, the application of his discovery went the furthest in fulfilling the Nobel criteria of demonstrating the greatest benefit to humankind in the preceding year. Unfortunately, the widespread use of DDT in peacetime came at great cost to animal life and with ongoing human health hazards posed by its persistence in the environment to this day.

A number of Nobel winners took credit for discoveries that were not particularly new. Robert Koch won for his introduction of rigorous and exacting laboratory methods in bacterial research. He was also widely credited with discovering the cholera bacteria, but *Vibrio cholerae* was first described in detail by Filippo Pacini in 1854 when Koch was a still a child.

Charles Laveran's prize for malaria research did not acknowledge that Patrick Manson had already proposed that malaria was transmitted by mosquitoes. When Ronald Ross traced the life cycle of the malaria parasite from birds to mosquitoes and back again, he was unknowingly following the path of a parasite unique to birds, one that does not infect humans at all. Grassi's Italian research team identified the correct parasite that infects humans and found the exact species of mosquito that carries it, but despite twenty-one nominations, he never won.

Alexander Fleming won for his 1928 discovery of penicillin, which had already been reported in the medical literature by junior Pasteur researcher Ernest Duchesne in 1897 when Fleming was still a teenager. Duchesne died in obscurity from TB while still a young man. Penicillin was rediscovered by the Costa Rican scientist Clodomiro Picado and described in his publications in 1923 and 1927.

The Nobel Prize was awarded to Christiaan Eijkman and Frederick Hopkins in 1929 for work they did much earlier on the disease beriberi.

Neither researcher had realized at the time that polished rice lacked a vital B vitamin. Beriberi had first been ascribed to a "missing factor" in white rice by British Army physician Samuel Hunter Christie in the 1830s. It was ultimately Umetaro Suzuki who determined in 1910 that polished rice specifically lacked vitamin B_1 (thiamine).

Gerhard Domagk won for creating the first sulfa antibiotic, but what he discovered in 1935 was found to be just a pro-drug, the body having to metabolize it to turn it into the active drug. The active drug was originally synthesized by an Austrian chemist in 1908 but neglected on a shelf and labeled as a dye because it had no known use in medicine at that time.

The Nobel Prize Committee carries out its deliberations in total secrecy, thus garnering somewhat of a mystique around divining the best in the world: the most original thinkers, the esteemed, first in discovery, individuals whose work most benefits humankind. Emil von Behring won for research on diphtheria. The Nobel Committee decided against the co-discoverers of the diphtheria toxin: Emile Roux (who had 115 nominations) and Alexandre Yersin (with 7 nominations). The committee also ignored the contributions of von Behring's laboratory colleague Kitasato Shibasaburo, codiscoverer of the diphtheria antitoxin with von Behring as well as the sole discoverer of tetanus toxin and tetanus toxoid. At that time, the diphtheria antitoxin and the toxoid were too strong to be made into a vaccine. It was not until the 1920s that Gaston Ramon, a microbiologist and veterinarian, developed a safer vaccine against diphtheria. A record 155 nominations (the most ever in the medicine category) were submitted for Ramon, but he did not win.

Malaria researchers Patrick Manson and Giovanni Battista Grassi had fifteen and twenty-one nominations, respectively, but the Nobel Committee recognized Robert Ross and Charles Laveran instead, even though both winners had some erroneous conclusions.

Fritz Schaudinn had four nominations for the 1905 prize for his discovery of the syphilis bacteria but never won. It was in Paul Ehrlich's lab that the discovery of the first chemotherapy for infectious disease was made. The actual discoverer of Salvarsan, Sahachiro Hata, had three nominations.

Emil Theodor Kocher won for thyroid surgery but was more famous for his meticulous operative techniques, which made surgery safe. He was a contemporary of Joseph Lister, who was the first to introduce the idea of aseptic techniques in surgery. Lister had fifteen nominations but was never recognized with a Nobel Prize.

Octave Gengou was always at the side of Nobel winner Jules Bordet in research on pertussis and the immune system activation of complement. Gengou submitted one of the 115 nominations for Bordet, but only one nomination came in for Gengou himself.

Corneille Jean François Heymans won for demonstrating that the heart was controlled by nerves emanating from the brain, drawing liberally from the original theories and illustrations of Fernando de Castro. It was actually de Castro who identified the exact nerve supply to the heart. He was never nominated.

At the time when Johannes Fibiger was given the prize for his erroneous conclusion that parasites caused stomach cancer, Katsusaburo Yamagiwa correctly demonstrated that chemicals could cause certain other cancers. He received seven nominations but never won.

The Nobel organization has nurtured excellent public relations to the point that mention of the prize evokes images of honor, justice, and morality. We can see shining examples of this in the characters of Nobel laureates such as Charles Sherrington, A. V. Hill, and Ramón y Cajal, but these are not necessarily consistent qualities among the winners.

Perhaps more than most other scientific pursuits, medicine has consistently been a product of its social environment. Nobel laureates contributed significantly to the tremendous advance of medicine in the first half of the twentieth century. Their stories reflect the personal and familial concerns common to their era as well as the particular local, national, and international pressures of the times. It is remarkable that some of them were able to persevere in their research and make revolutionary discoveries at all. They were not always correct in their scientific theories or fair in their dealings with colleagues and coworkers, nor did they always have decent intentions toward humankind. Likewise, the Nobel Committee members selecting the winners were subject to human shortcomings and external influences. It would be astute for the nonmedical and medical reader alike to be cautious about assigning unwarranted authority to Nobel Prize winners. This seems easy in retrospect but can be a challenge in the moment, especially when faced with the nearly universal high regard enjoyed by the Nobel Prize.

Readers are invited to delve into the next fifty years of the Nobel Prize in Medicine in part 2 of this series.

Points of Interest

Fiction and Philosophical Works by Nobel Laureates

Antônio Egas Moniz

A vida sexual [The sexual life]. Lisbon: Casa Ventura Abrantes, 1922.
História das cartas de jogar [History of playing cards]. Lisbon: Apenas Libros, 1942.
O Padre Faria na história do hipnotismo [Father Faria in the history of hypnotism]. Lisbon: Libânio da Silva, 1925.

Santiago Ramón y Cajal

Vacation Stories—Five Science Fiction Tales. Champaign: University of Illinois Press, 2006.
Advice for a Young Investigator. Cambridge, MA: MIT Press, 2004.
Life in the Year 6000: A Fantasy Dream. Scotts Valley, CA: CreateSpace, 2017.
The Psychology of Don Quixote and the Quixotic Ideal. Scotts Valley, CA: CreateSpace, 2016.

Charles Jules Henri Nicolle

The Pâtissier de Bellone. Paris: Calmann-Lévy, 1913.
The Leaves of Sagittarius. Paris: Calmann-Levy, 1920.
La Narquoise. Paris: Calmann-Levy, 1922.
The Plaisirs of Trouble Menus. Paris: Rieder, 1924.
Marmouse and His Guests. Paris: Rieder, 1927.
The Two Larrons. Paris: Calmann-Levy, 1929.

Alexis Carrell

Man, The Unknown. Garden City, NY: Halycon House, 1938.

Charles Richet
L'homme stupide [Idiot man]. Scotts Valley: CreateSpace, 2016.

Charles Scott Sherrington
Man on His Nature. Cambridge: University Press, 1940.

Albert Szent-Györgyi
The Crazy Ape. New York: Philosophical Library, 1970.
What's Next. New York: Philosophical Library, 1971.
Science, Ethics and Politics. New York: Vantage Press, 1963.

Hermann Joseph Muller
Out of the Night: A Biologist's View of the Future. New York: Vanguard, 1955.

Eugenicists

Egas Moniz
Hermann Joseph Muller
Thomas Hunt Morgan
Alexis Carrel
Charles Richet

Nazis

Julius Wagner-Jauregg
Charles Richet
Communist Affiliations
Ivan Pavlov
Hermann Joseph Muller

Mystics

Alexis Carrel
Charles Richet
Associated with Rockefeller Institute
Karl Landsteiner
Herbert Spencer Gasser
Alexis Carrel

Associated with Pasteur Institute

Charles Louis Henri Nicolle
Charles Louis Alphonse Laveran
Ilya Ilyich Mechnikov
Jules Bordet

Associated with Kaiser Wilhelm Institute

Otto Fritz Meyerhof
Otto Heinrich Warburg
Hermann Joseph Muller
Female Winners
Gerty Cori

Laureates from High-Chocolate-Consuming Nations

Statistically, more Nobel Prize winners come from high-chocolate-consuming nations. Twenty-three out of fifty-seven total winners in medicine may have had chocolate to thank for their success.

Highest per capita chocolate consumption/winners in medicine:

1. Switzerland (Walter Rudolf Hess, Paul Hermann Müller, Emil Theodor Kocher)
2. Germany (Ernst Boris Chain, Otto Loewi, Otto Heinrich Warburg, Otto Fritz Meyerhof, Robert Koch)
3. UK (Henry Hallett Dale, Charles Scott Sherrington, Edgar Adrian, Frederick Gowland Hopkins, Archibald Vivian Hill)
4. Norway (0)
5. Ireland (0)
6. Denmark (Henrik Dam, Johannes Fibiger, August Krogh)
7. Austria (Julius Wagner-Jauregg, Róbert Bárány)
8. Finland (0)
9. Sweden (Allvar Gullstrand)
10. France (Charles Louis Henri Nicolle, Charles Richet, Alexis Carrel, Charles Louis Alphonse Laveran)

References

Introduction

Fant, Kenne, *Alfred Nobel: A Biography*, trans. Marianne Ruuth (New York: Arcade Publishing, 2014).

Feldman, Burton, *The Nobel Prize: A History of Genius, Controversy, and Prestige* (New York: Arcade Publishing, 2001).

Kozelsky, Mara, "The Crimean War, 1853–56," *Kritika: Explorations in Russian and Eurasian History* 13, no. 4 (Fall 2012).

Lemmel, Birgitta, "Alfred Nobel–St. Petersburg, 1842–1863," Nobel Media AB 2014, http://www.nobelprize.org/alfred_nobel/biographical/articles/russia/.

The Nobel Prize, "Prize Amount and Market Value of Invested Capital Converted into 2014 Year's Monetary Value," updated December 2014; http://www.nobelprize.org/nobel_prizes/about/amounts/prize_amounts_15.pdf.

Chapter 1

Humprhies, Suzanne and Roman Bystrianyk, *Dissolving Illusions: Disease, Vaccines, and the Forgotten History* (Scotts Valley, CA: CreateSpace, 2013).

Linton, Derek S., *Emil Von Behring: Infectious Disease, Immunology, Serum Therapy, Memoirs of the American Philosophical Society* (Philadelphia: American Philosophical Society, 2005).

New York Times, "The Antitoxin Scandal in St. Louis," February 14, 1902.

New York Times, "The Population of the Island of Capri Is Indignantly Protesting," November 5, 1903.

New York Times, "Behring Denies He Is Insane," February 8, 1907.

Von Behring, Emil and Charles Bolduan, *Suppression of Tuberculosis, Together with Observations Concerning Phthisiogenesis in Man and Animals and Suggestions Concerning the Hygiene of Cow Stables and the Production of Milk for Infant Feeding, with Special Reference to Tuberculosis*, reproduction (San Bernardino, CA: Ulan Publishing, 2011).

Chapter 2

Chernin, E., "Sir Ronald Ross vs. Sir Patrick Manson: A Matter of Libel," *Journal of the History of Medicine and Allied Sciences* 43, no. 3 (1988): 262–74.

Lerner, K. L. and Brenda Wilmoth Lerner, *World of Anatomy and Physiology* (Detroit, MI: Gale, 2002).

Nye, Edwin R. and Mary E. Gibson, *Ronald Ross: Malariologist and Polymath, a Biography* (London: Palgrave Macmillan, 1997).

Ross, Ronald, Sr., *Memories of Sir Patrick Manson* (London: Harrison & Sons, 1930).

Chapter 3

Daniel, Thomas M., "The Impact of Tuberculosis on Civilization," *Infectious Disease Clinics of North America* 18, no. 1 (2004): 157–65.

Gøtzsche, P. C., "Niels Finsen's Treatment for Lupus Vulgaris," *Journal of the Royal Society of Medicine* 104, no. 1 (2011): 41–42.

Grzybowski, A. and K. Pietrzak, "From Patient to Discoverer—Niels Ryberg Finsen (1860–1904)—The Founder of Phototherapy in Dermatology," *Clinical Dermatology* 30, no. 4 (2012): 451–55.

Liu, P. T., S. Stenger, L. Wenzel et al., "Toll-Like Receptor Triggering of a Vitamin D-Mediated Human Antimicrobial Response," *Science* 311 (2006): 1770–73.

Chapter 4

Babkin, P., *Pavlov: A Biography* (Chicago: University of Chicago Press, 1949).

Lionni, Paolo, *The Leipzig Connection* (Sheridan, OR: Heron Books, 1993).

Meerloo, Joost A. M., *The Rape of the Mind: The Psychology of Thought Control, Menticide, and Brainwashing* (New York: World Publishing Company, 1956).

Pavlov, Ivan Petrovitch, "Excerpts from Lectures on Conditioned Reflexes, Volume II: Conditioned Reflexes and Psychiatry," chapter 46, "Experimental Neuroses" (read in German at the First International Neurological Congress, Berne, September 3, 1931).

Todes, Daniel Philip, *Ivan Pavlov: A Russian Life in Science* (New York: Oxford University Press, 2014).

Windholz, George, "Pavlov's Religious Orientation," *Journal for the Scientific Study of Religion* 25, no. 3 (September 1986): 320–27.

Chapter 5

Bentivoglio, Marina and Paolo Pacini, "Filippo Pacini: A Determined Observer," *Brain Research Bulletin* 38, no. 2 (1995): 161–65.

Brock, Thomas D., *Robert Koch: A Life in Medicine and Bacteriology* (Washington, DC: ASM Press, 1999).

Goetz,Thomas, *The Remedy: Robert Koch, Arthur Conan Doyle, and the Quest to Cure Tuberculosis* (New York: Gotham, 2015).

Howard-Jones, N., "Robert Koch and the Cholera Vibrio: A Centenary," *British Medical Journal* 288, no. 6414 (February 4, 1984): 379–81.

Chapter 6

Cimino, Guido, "Reticular Theory versus Neuron Theory in the Work of Camillo Golgi," *Physis: Rivista internazionale di storia della scienza* 36, no. 2 (1999): 431–72.

Golgi, Camillo, *Annotazioni intorno all'istologia dei reni dell'uomo e di altri mammifieri e sull'istogenesi dei canalicoli oriniferi*, vol. 5 (Rome: Rendiconti della Reale Accademia dei Lincei, 1889), 545–57.

Pannese, Ennio, "The Golgi Stain: Invention, Diffusion and Impact on Neurosciences" *Journal of the History of the Neurosciences* 8, no. 2 (August 1999): 132–40.

Ramón y Cajal, Santiago, *Vacation Stories: Five Science Fiction Tales*, trans. Laura Otis (Champaign: University of Illinois Press, 2001).

Tzitsikas, Helene, *Santiago Ramón y Cajal: Obra literaria*, vol. 53, Colección Studium (Mexico City: De Andrea, 1965).

Zinn, Howard, Mike Konopacki, and Paul Buhle, A *People's History of American Empire: The American Empire Project* (New York: Metropolitan Books, 2008).

Chapter 6

Centers for Disease Control (CDC), "Laveran and the Discovery of the Malaria Parasite," December 14, 2007, https://www.cdc.gov/malaria/about/history/laveran.html.

Kakkilaya, Bevinje Srinivas, "Charles Louis Alphonse Laveran (1845–1922)," Malaria Site, February 25, 2015, http://www.malariasite.com//?s=laveran.

The Nobel Prize, "Alphonse Laveran," December 14, 2007, http://nobelprize.org/nobel_prizes/medicine/laureates/1907/laveran-bio.html.

Nye, Edwin R., "Alphonse Laveran (1845–1922): Discoverer of the Malarial Parasite and Nobel Laureate, 1907," *Journal of Medical Biography* 10, no. 2 (June 2002): 81–87.

Chapter 8

Hirsch, James G., "Immunity to Infectious Diseases: Review of Some Concepts of Metchnikoff," *Bacteriological Reviews* 23, no. 2 (June 1959): 48–60.

Kaufmann, Stefan H. E., "Immunology's Foundation: The 100-Year Anniversary of the Nobel Prize to Paul Ehrlich and Elie Metchnikoff," *Nature Immunology* 9 (2008): 705–12.

Leyden, John G., "From Nobel Prize to Courthouse Battle; Paul Ehrlich's 'Wonder Drug' for Syphilis Won Him Acclaim but Also Led Critics to Hound Him," *Washington Post*, July 27, 1999.

Metchnikoff, Elie, *The Nature of Man, Studies in Optimistic Philosophy* (New York: G. P. Putnam's Sons, 1903; London: Forgotten Books, 2012).

New York Times, "Prof. Ehrlich In Libel Suit; Denies Newspaper Man's Charge That Salvarsan Is Dangerous," June 9, 1914.

Weissmann, George, "Dr. Ehrlich and Dr. Atomic: Beauty vs. Horror in Science," *FASEB Journal* 23 (January 2009).

Chapter 9

Gautschi, Oliver P. and Gerhard Hildebrandt, "Emil Theodor Kocher (25/8/1841–27/7/1917)—A Swiss (Neuro-)Surgeon and Nobel Prize Winner," *British Journal of Neurosurgery* 23, no. 3 (July 2009): 234–36.

Kocher, Emil Theodor, "Über Kropfexstirpation und ihre Folgen [On goitre removal, and its consequences]," *Archiv für Klinische Chirurgie* 29 (1883): 254–337.

Messerli, Franz H., "Chocolate Consumption, Cognitive Function, and Nobel Laureates," *New England Journal of Medicine* 367, no. 16 (2012): 1562–64.

Ravin, James G., "Gullstrand, Einstein, and the Nobel Prize," *Archives of Ophthalmology* 117 (May 1999).

Slater, Stefan, "The Discovery of Thyroid Replacement Therapy, part 3: A Complete Transformation," *Journal of the Royal Society of Medicine* 104, no. 3 (March 2011): 100–106.

Chapter 10

Carrel, Alexis, *The Voyage to Lourdes* (New York: Harper & Brothers, 1950; New Hope, KY: Real View Books, 1994). Originally published posthumously; the 1994 edition with new preface by Stanley Jaki corrects errors in other versions.

Reggiani, Andrés Horacio, *God's Eugenicist: Alexis Carrel and the Sociobiology of Decline* (New York: Berghahn Books, 2007).

Witkowski, Jan A., "Dr. Carrel's Immortal Cells," *Medical History* 2 (April 24, 1980): 129–42.

Chapter 11

Ardaillou, Raymond A. and Pierre M Ronco, "Obituary: Gabriel Richet (1916–2014)," *Kidney International* 87, no. 1 (2015): 3–4.

Bonds, Rana S.and Brent C. Kelly, "Severe Serum Sickness After H1N1 Influenza Vaccination," *American Journal of the Medical Sciences* 345, no. 5 (May 2013): 412–13.

Krimbas, C., "Eugenics in Europe," *International Encyclopedia of Social and Behavioral Sciences*, 1st ed., ed. Neil J. Smelser and Paul B. Baltes (Oxford: Elsevier, 2001), s.v. "eugenics in Europe."

New York Times, "Clashes With Houdini: Youth Who Reads Through Metal Charges Trickery at Tests," May 9, 1924.

Richet, Charles R., *Our Sixth Sense*, trans. Fred Rothwell (London: Rider & Company, 1928).
——— [Charles Epheyre, pseud.], "Professor Bakermann's Microbe, A Tale of the Future," trans. Brian Stableford, in *The Supreme Progress* (Los Angeles: Black Coat Press, 2011).
——— and Raoul Brunel, *Circé: drame en deux actes* (Paris: Choudens, 1903).
———, Charles Epheyre et Octave Houdaille. *Soeur Marthe, drame lyrique en 2 parties, 3 actes et 5 tableaux* (Paris: Ollendorff, 1898).
Schneider, William H., *Quality and Quantity: The Quest for Biological Regeneration in Twentieth-Century France* (Cambridge: Cambridge University Press, 1990).
Society for Psychical Research, "Past Presidents," https://www.spr.ac.uk/about/past-presidents.
Wolf, Stewart, *Brain, Mind, and Medicine: Charles Richet and the Origins of Physiological Psychology* (Piscataway, NJ: Transaction Publishers, 1993).

Chapter 12

Carey, Michael E., "Cushing and the Treatment of Brain Wounds During World War I, Historical Vignette," *Journal of Neurosurgery* 114, no. 6 (June 2011): 1495–1501.
Grady Tim, *The German-Jewish Soldiers of the First World War in History and Memory* (Liverpool: Liverpool University Press, 2012.
Kassemi, Mohammad, Dimitri Deserranno, and John Oas, "Effect of Gravity on the Caloric Stimulation of the Inner Ear," *Annals of the New York Academy of Sciences* 1027 no. 1 (November 2004): 360–70.
Mudry, A., "Neurological Stamp: Robert Barany (1876–1936)," Journal of Neurology, Neurosurgery & Psychiatry 68 (2000): 507.
Wilford, John Noble, "Space Test Jolts Inner-Ear Theory," *New York Times*, December 7, 1983.

Chapter 13

Encyclopedia.com, s.v. "Bordet, Jules (1870–1961)," in "World of Microbiology and Immunology," last modified June 21, 2015, https://www.encyclopedia.com/people/medicine/medicine-biographies/jules-bordet.
Geier, Mark R. and David A Geier, "The True Story of Pertussis Vaccine: A Sordid Legacy?" *Journal of the History of Medicine and Allied Sciences* 57, no. 3 (July 2002): 249–84.
Humphries, Suzanne and Roman Bystrianyk, *Dissolving Illusions: Disease, Vaccines, and the Forgotten History* (Scotts Valley, CA: CreateSpace, 2013). For the effect of the vaccine on whooping cough, see the graph at http://www.dissolvingillusions.com/graphs/#12.

Lewis, Sinclair, *Arrowsmith* (New York: Harcourt Brace & Company, 1925). Silverstein, Arthur M., *Paul Ehrlich's Receptor Immunology: The Magnificent Obsession* (Cambridge, MA: Academic Press, 2001).

Ligon, B. Lee, "Jules Bordet: Pioneer Researcher in Immunology and Pertussis (1870–1961)," *Seminars in Pediatric Infectious Diseases* 9, no. 2 (April 9, 1998): 163–67.

Oakley, C. L., "Jules Jean Baptiste Vincent Bordet, 1870–1961," *Biographical Memoirs of Fellows of the Royal Society* 8 (1962): 18.

Summer, William C., "The Strange History of Phage Therapy," *Bacteriophage* 2, no. 2 (April 1, 2012): 130–33.

Chapter 14

American Medical Association, "August Krogh (1874–1949) The Physiologist's Physiologist," *Journal of the American Medical Association* 199, no. 7 (1967): 496–97.

Krogh, August, "A Contribution to the Physiology of the Capillaries (Nobel Lecture, December 11, 1920)," *Nobel Lectures, Physiology or Medicine* 1901–1921 (Amsterdam: Elsevier. 1967).

———, *The Anatomy and Physiology of Capillaries* (New Haven, CT: Yale University Press, 1922).

Chapter 15

Bassett, David R., Jr., "Scientific Contributions of A. V. Hill: Exercise Physiology Pioneer," *Journal of Applied Physiology* 93, no. 5 (November 1, 2002): 1567–82.

Buderi, Robert, *The Invention That Changed the World: How a Small Group of Radar Pioneers Won the Second World War and Launched a Technological Revolution* (New York: Simon & Schuster, 1996).

Hill, Archibald Vivian, *The Ethical Dilemma of Science and Other Writings* (New York: Rockefeller University Press, 1960).

———, "International Status and Obligations of Science. Letter to the Editor (Reply)," *Nature* 133, no. 3356 (1934): 290.

———, *Trails and Trials in Physiology: A Bibliography, 1909–1964; with Reviews of Certain Topics and Methods and a Reconnaissance for Further Research* (Baltimore: Williams and Wilkins, 1965)

Katz, Bernard, "Archibald Vivian Hill. 26 September 1886–3 June 1977," *Biographical Memoirs of Fellows of the Royal Society* 24 (November 1978): 71–149.

Pearson, Karl, "On the Inheritance of Mental Disease," *Annals of Eugenics* 4, nos. 3–4 (1931): 362–80.

———, "On a New Theory of Progressive Evolution," *Annals of Eugenics* 4, nos. 1–2 (1930): 1–40.

Pyke, D., "Contributions by German Émigrés to British Medical Science," *Quarterly Journal of Medicine: An International Journal of Medicine* 93, no. 7 (July 2000): 487–95.

Squire Larry R., ed., *The History of Neuroscience in Autobiography, Volume 1* (Washington, DC: Society for Neuroscience, 1996).

Stark, Johannes, "International Status and Obligations of Science (Letter to the Editor)," *Nature* 133, no. 3356 (1934): 290.

Szöllösi-Janze, Margit, *Science in the Third Reich* (Oxford: Berg, 2001).

Chapter 16

"Bhaishagykni, Charm to Secure Perfect Health," Atharva Veda 2.32, trans. Maurice Bloomfield, in *Sacred Books of the East*, vol. 42 (Oxford: Oxford University Press, 1924).

"Dr. Fred Banting, Nobel Laureate," reported by Rae Corelli, Sunday Morning, broadcast November 15, 1981, Canadian Broadcasting Company Radio Archives.

Papaspyros, N. S., The History of Diabetes Mellitus (Stuttgart: Georg Thieme Verlag, 1964).

Saravanan, Ramalingam and Leelavinothan Pari, "Antidiabetic Effect of Diasulin, an Herbal Drug, on Blood Glucose, Plasma Insulin and Hepatic Enzymes of Glucose Metabolism in Hyperglycaemic Rats," *Diabetes Obesity and Metabolism* 6, no. 4 (July 2004): 286–92.

Chapter 17

Foster, Ruth, *Take Five Minutes: Fascinating Facts and Stories for Reading and Critical Thinking (Take 5 Minutes)* (Garden Grove, CA: Teacher Created Resources, 2001).

Harvard Crimson, "America to Lead World in Scientific Research, Professor Einthoven Impressed with Boston Laboratories—Noted for Study of Human Electricity," November 1, 1924, http://www.thecrimson.com/article/1924/11/1/america-to-lead-world-in-scientific/.

Lama, Alexis, "Einthoven. El hombre y su invento," [The Man and His Invention] *Revista médica de Chile* 132, no. 2 (March 2004): 260–64.

Myers, Morton A., *Happy Accidents: Serendipity in Major Medical Breakthroughs in the Twentieth Century* (New York: Skyhorse Publishing, 2011).

Rivera-Ruis, Moises, Christian Cajavilca, and Joseph Varon, "Einthoven's String Galvanometer: The First Electrocardiograph," *Texas Heart Institute* 35, no. 2 (2008): 174–78.

Silverman, Mark E. and J. Willis Hurst, "Willem Einthoven—The Father of Electrocardiography," *Clinical Cardiology* 15, no. 10 (October 1992): 785–87.

Chapter 18

Fibiger, Johannes, "On *Spiroptera carcinomata* and Their Relation to True Malignant Tumors; With Some Remarks on Cancer Age," *Journal of Cancer Research* 4, no. 4 (October 1919): 367–87.

"Katsusaburo Yamagiwa," *CA: A Cancer Journal for Clinicians* 27, no. 3 (May–June 1977): 172–73.

Stolley, Paul D. and Tamar Lasky, "Johannes Fibiger and His Nobel Prize for the Hypothesis That a Worm Causes Stomach Cancer," *Annals of Internal Medicine* 116, no. 9 (May 1992): 765–69.

Rous, Peyton, "Sarcoma of the Fowl Transmissible by an Agent Separable from the Tumor Cells," *Journal of Experimental Medicine* 13, no. 4 (April 1, 1911): 397–411.

Yamagiwa, Katsusaburo and Koichi Ichikawa, "Experimental Study of the Pathogenesis of Carcinoma," *CA: A Cancer Journal for Clinicians* 27, no. 3 (May–June 1977): 174–81.

Chapter 19

Allerberger, F., "Julius Wagner-Jauregg (1857-1940)," *Journal of Neurological Psychiatry* 62, no. 3 (1997): 221.

Brabin, Bernard, "Malaria's Contribution to World War One—The Unexpected Adversary," *Malaria Journal* 13 (2014): 497.

Chapman, Clare, "Austrians Stunned by Nobel Prize-Winner's Nazi Ideology," broadcast on *Scotland on Sunday*, January 25, 2004.

Cormier, Loretta A., *The Ten-Thousand Year Fever: Rethinking Human and Wild-Primate Malarias* (Walnut Creek, CA: Left Coast Press, 2011).

Hoffman, Edward, *The Drive for Self: Alfred Adler and the Founding of Individual Psychology* (New York: Addison-Wesley Publishing, 1994).

Idro, Richard et al., "Cerebral Malaria Is Associated with Long-Term Mental Health Disorders: A Cross Sectional Survey of a Long-Term Cohort," *Malaria Journal* 15 (March 2016): 184.

Rudolf, G. De M. , "Recent Advances in Therapeutic (Induced) Malaria," *Journal of Neurological Psychopathology* 16, no. 63 (January 1936): 497.

Tsay, Cynthia J., "Julius Wagner-Jauregg and the Legacy of Malarial Therapy for the Treatment of General Paresis of the Insane," *Yale Journal of Biological Medicine* 86, no. 2 (June 2013): 245–54..

Varney, Nils R. et al., "Neuropsychiatric Sequelae of Cerebral Malaria in Vietnam Veterans," *Journal of Nervous and Mental Disease* 185, no. 11 (November 1997).

Wagner-Jauregg, Julius, "The Treatment of Dementia Paralytica by Malaria Inoculation (Nobel Lecture, December 13, 1927)," *Nobel Lectures in Physiology or Medicine, 1922–1941* (Amsterdam: Elsevier, 1965).

Chapter 20

Conseil, Ernest, "Le Typhus exanthématique en Tunisie pendant l'année 1909," [The Typhus Exanthematicus in Tunisia During the Year 1909], *Archives de l'Institut Pasteur de Tunis*, March 1910.

Gross, L., "How Charles Nicolle of the Pasteur Institute Discovered That Epidemic Typhus Is Transmitted by Lice: Reminiscences from My Years at the Pasteur Institute in Paris," *Proceeding of the National Academy of Sciences of the United States of America* 93, no. 20 (October 1, 1996): 10539-40.

Nicolle, Charles, *Le pâtissier de Bellone* (Paris: Calmann-Lévy, 1913).

———, *The Leaves of Sagittarius* (Paris: Calmann-Lévy, 1920).

———, *La narquoise* (Paris: Calmann-Lévy, 1922).

———, *The Plaisirs of Trouble Menus* (Paris: Rieder, 1924)

———, *Marmouse and His Guests* (Paris: Rieder, 1927)

———, *The Two Larrons* (Paris: Calmann-Lévy, 1929).

Pelis, Kim, Charles Nicolle, *Pasteur's Imperial Missionary: Typhus and Tunisia* (Rochester, NY: University of Rochester Press, 2006).

Schultz, Myron G. and David M. Morens, "Charles-Jules-Henri Nicolle," *Emerging Infectious Diseases* 15, no. 9 (September 2009): 1520–22, https://www.nc.cdc.gov/eid/pdfs/vol15no9_pdf-version.pdf.

Chapter 21

Hopkins, Frederick, "The Earlier History of Vitamin Research (Nobel Lecture, December 11, 1929)," *Nobel Lectures, Physiology or Medicine, 1922–1941* (Amsterdam: Elsevier, 1965).

Price, Catherine, *Vitamania: Our Obsessive Quest for Nutritional Perfection* (New York: Penguin Press, 2015).

Sugiyama, Yoshifumi and Akihiro Seita, "Kanehiro Takaki and the Control of Beriberi in the Japanese Navy." *Journal of the Royal Society of Medicine* 106, no. 8 (2013): 332–34.

Chapter 22

"Dr. Landsteiner Sues to Escape Being Labelled Jew," Archives of the Jewish Telegraphic Agency, April 6, 1937, www.jta.org/1937/04/06/archive/dr-landsteiner-sues-to-escape-being-labelled-Jew.

Kantha, S. S., "Is Karl Landsteiner the Einstein of the Biomedical Sciences?" *Medical Hypotheses* 44, no. 4 (April 1995): 254–56.

———"The Blood Revolution Initiated by the Famous Footnote of Karl Landsteiner's 1900 Paper," *Ceylon Medical Journal* 40, no. 3 (September 1995): 123–25.

Owen, Ray, "Karl Landsteiner and the First Human Marker Locus," *Genetics* 155, no. 3 (July 1, 2000): 995–98.

Simmons, John Galbraith, *Scientific 100: A Ranking of the Most Influential Scientists, Past and Present* (New York: Citadel Press, 2000).

Tan, Siang Yong and Connor Graham, "Karl Landsteiner (1868–1943): Originator of ABO Blood Classification," *Singapore Medical Journal* 54, no. 5 (2013): 243–44.

Chapter 23

Krebs, Hans Adolf and Roswitha Schmid, *Otto Warburg: Zellphysiologe, Biochemiker, Mediziner: 1883–1970*, Grosse Naturforscher Band 41 (Stuttgart: Wissenschaftliche Verlagsgesellschaft, 1979).

Medawar, Jean and David Pyke, Hit*ler's Gift: The True Story of Scientists Expelled by the Nazi Regime* (New York: Arcade Publishing, 2012).

Otto, Angela M., "Warburg Effect(s)—A Biographical Sketch of Otto Warburg and His Impacts on Tumor Metabolism," *Cancer and Metabolism* 4, no. 5 (2016).

Vennesland, Birgit, review of *New Methods of Cell Physiology*, by Otto Heinrich Wagner, *Perspectives in Biology and Medicine* 6, no. 3 (Spring 1963): 385–88.

Warburg, Otto H., "On the Origin of Cancer Cells: *Science* 123, no. 3191 (February 24, 1956): 309–14.

———, "The Prime Cause and Prevention of Cancer," revised lecture at the meeting of the Nobel Laureates, June 30, 1966.

——— Franz Wind, and Erwin Negelein, "The Metabolism of Tumors in the Body," *Journal of General Physiology* 8, no. 6 (March 7, 1927): 519–30.

Weisz, George M., "Dr. Otto Heinrich Warburg—Survivor of Ethical Storms," *Rambam Maimonides Medical Journal* 6, no. 1 (January 2015): e0008.

Chapter 24

Eccles, John C., *How the Self Controls Its Brain* (Berlin: Springer-Verlag, 1994).

——— and William C. Gibson, *Sherrington, His Life and Thought* (New York: Springer International, 1979).

Sherrington, Charles, "Inhibition as a Coordinative Factor (Nobel Lecture, December 12, 1932)," *Nobel Lectures, Physiology or Medicine, 1922–1941* (Amsterdam: Elsevier, 1965).

Stuart, Douglas G. et al., "Sir Charles S. Sherrington: Humanist, Mentor, and Movement Neuroscientist, *Classics in Movement Science*, ed. Mark L. Latash and Vladimir M. Zatsiorsky (Champaign, IL: Human Kinetics, 2001).

Chapter 25

Berrington de González, Amy et al., "Projected Cancer Risks from Computed Tomographic Scans Performed in the United States in 2007," *Archives of Internal Medicine* 169, no. 22 (2009): 2071–77.

Institute of Medicine, *Breast Cancer and the Environment: A Life Course Approach* (Washington, DC: Institute of Medicine of the National Academies Press, 2011), http://www.nationalacademies.org/hmd/Reports/2011/Breast-Cancer-and-the-Environment-A-Life-Course-Approach.aspx.

Chapter 26

Castle, W. B., "George Richards Minot," *Biographical Memoirs of the National Academy of Sciences*, vol. 45 (Washington, DC: National Academies Press, 1974): 353–401.

Miller, Leon, "George Hoyt Whipple," *Biographical Memoirs of the National Academy of Sciences*, vol. 66 (Washington, DC: National Academies Press, 1995): 370–93.

Whipple, George Hoyt and Frieda Robscheit-Robbins, "Blood Regeneration in Severe Anemia: II. Favorable Influence of Liver, Heart and Skeletal Muscle in Diet," *American Journal of Physiology* 72 (1925): 408–18.

Chapter 27

Hamburger, Viktor, *The Heritage of Experimental Embryology: Hans Spemann and the Organizer* (New York: Oxford University Press, 1988).

Harrington, Anne, *Reenchanted Science: Holism in German Culture from Wilhelm II to Hitler* (Princeton, NJ: Princeton University Press, 1999)

Ribatti, D., "The Chemical Nature of the Factor Responsible for Embryonic Induction: An Historical Overview," *Organogenesis* 10, no. 1 (January 1, 2014): 38–43.

Spemann, Hans, *Embryonic Development and Induction* (New York: Hafner Publishing Company, 1967).

Wellner, K., "Hans Spemann (1869–1941)," *Embryo Project Encyclopedia* (Tempe: Arizona State University, Center for Biology and Society, June 15, 2010), http://embryo.asu.edu/pages/hans-spemann-1869-1941.

Chapter 28

Campbell, G., "Cotransmission," *Annual Review of Pharmacological Toxicology* 27 (1987): 51–70.

Feldberg, W. S., "Henry Hallett Dale, 1875–1968," *Biographical Memoirs of Fellows of the Royal Society* 16 (November 1970): 77–174.

Hurst, J. Willis, W. Bruce Fye, and Heinz-Gerd Zimmer, "Otto Loewi and the Chemical Transmission of Vagus Stimulation in the Heart," *Clinical Cardiology* 29, no. 3 (March 2006): 135–36.

Loewi, Otto, "The Chemical Transmission of Nerve Action (Nobel Lecture, December 12, 1936)," *Nobel Lectures, Physiology or Medicine, 1922–1941* (Amsterdam: Elsevier, 1965).

McCoy, Alli N. and Yong Siang Tan, "Otto Loewi (1873–1961): Dreamer and Nobel Laureate," *Singapore Medical Journal* 55, no. 1 (January 2014): 304.

Sabbatini, Renato M. E., "Neurons and Synapses: The History of Its Discovery," pt. 5, "Chemical Transmission," *Brain and Mind* 17 (April–July 2003), http://www.cerebromente.org.br/n17/history/neurons5_i.htm.

Tansey, E. M., "The Early Education of a Nobel Laureate: Henry Dale's Schooldays," *Royal Society Journal of the History of Science* 65, no. 4 (December 20, 2011): 379–91.

Valenstein, Elliot S., *The War of the Soups sand the Sparks: The Discovery of Neurotransmitters and the Dispute over How Nerves Communicate* (New York: Columbia University Press, 2006).

York, K. George, III, "Otto Loewi: Dream Inspires a Nobel-Winning Experiment on Neurotransmission," *Neurology Today* 4, no. 12 (December 2004): 54–55.

Chapter 29

Hargittai, Balazs and Istvan Hargittai, ed., "The Wit and Wisdom of Albert Szent-Györgyi: A Recollection," *Culture of Chemistry: The Best Articles on the Human Side of Twentieth-Century Chemistry* (New York: Springer, 1988).

Swarup, A., S. Stuchly, and A. Surowiec, "Dielectric Properties of Mouse MCA1 Fibrosarcoma at Different Stages of Development," *Bioelectromagnetics* 12, no. 1 (1991): 1–8.

Szent-Györgyi, Albert, "Lost in the Twentieth Century," *Annual Review of Biochemistry* 32 (1963): 1–15.

———, *Science, Ethics and Politics* (New York: Vantage Press, 1963).

———, *Some Misplaced Ideas on Democracy*, audio CD (Madison, CT: Jeffrey Norton Publishing, 2006), https://www.amazon.in/Misplaced-Ideas-Democracy-Albert-Szent-Gyorgyi/dp/1579703682/.

———, *What Next?* (New York: Philosophical Library, 1971).

US National Library of Medicine, "The Albert Szent-Györgyi Papers," Profiles in Science (Bethesda, MD: US National Library of Medicine), https://profiles.nlm.nih.gov/spotlight/wg/feature/biographical-overview.

Chapter 30

De Castro, Fernando, "The Discovery of the Sensory Nature of the Carotid Bodies," *Advances in Experimental Medicine and Biology* 648 (2009): 1–18.

———, "Towards the Sensory Nature of the Carotid Body: Hering, de Castro and Heymans," *Frontiers in Neuroanatomy* 3 (2009): 23.

Heymans, Jean François and Corneille Jean François Heymans, "Recherches physiologiques et pharmacodynamiques sur la tête isolée du chien," *Archives internationale de pharmacodynamique et de thérapie* 32 (1926): 1–33.

Chapter 31

Colebrook, L., "Gerhard Domagk, 1895–1964," *Biographical Memoirs of Fellows of the Royal Society* 10 (1964): 38–50.

Hager, Thomas, *The Demon Under the Microscope: From Battlefield Hospitals to Nazi Labs, One Doctor's Heroic Search for the World's First Miracle Drug* (New York: Harmony Books, 2006).

Kiefer, David M., "Miracle Medicines: The Advent of Sulfa Drugs in the Mid-1990s Gave Physicians a Powerful Weapon," *Today's Chemist at Work* (June 2001), http://pubsapp.acs.org/subscribe/archive/tcaw/10/i06/html/06chemch.html.

Lesch, John E., *How the Sulfa Drugs Transformed Medicine* (New York: Oxford University Press, 2006).

New York Times, "Death Drug Hunt Covered 15 States; Wallace Reveals How Federal Agents Traced Elixir to Halt Fatalities," November 26, 1937.

Richards, Ira Steven and Marie Bourgeois, *Principles and Practice of Toxicology in Public Health* (Burlington, MA: Jones and Bartlett Learning, 2013).

Chapter 32

Eissenberg, Joel C. and Enrico Di Cera, "Edward Adelbert Doisy, In Vitro Veritas: Ninety Years of Biochemistry at St. Louis University," *Missouri Medicine* 110, no. 4 (July/August 2013): 297–301.

Okamoto, Hiroshi, "Vitamin K and Rheumatoid Arthritis," *IUBMB Life* 60, no. 6 (June 2008): 355–61.

Zetterström, Rolf, "H.C.P. Dam (1895–1976) and E. A. Doisy (1893–1986): The Discovery of Antihaemorrhagic Vitamin and Its Impact on Neonatal Health," *Acta Paediatrica* 95, no. 6 (July 2006): 642–44.

Chapter 33

Chase, Merrill W. and Carlton C. Hunt, "Herbert Spencer Gasser (1888–1963): A Biographical Memoir" (Washington, DC: National Academies Press, 1995).

Davis, Hallowell, "Joseph Erlander (1874–1965): A Biographical Memoir" (Washington, DC: National Academies Press, 1970).

Encyclopedia Britannica, 11th ed. (1911), s.v. "Giovanni Aldini."

McComas, Alan J., *Galvani's Spark: The Story of the Nerve Impulse* (New York: Oxford University Press, 2011).

Parent, André, "Giovanni Aldini: From Animal Electricity to Human Brain Stimulation," *Canadian Journal of Neurological Science* 31, no. 4 (November 2004): 576–84.

Chapter 34

Palmer, Steven and Ivan Molina, ed., *The Costa Rica Reader: History, Culture, Politics* (Durham, NC: Duke University Press, 2004).

Chain, E. B. , "Contributions from Chemical Microbiology to Therapeutic Medicine," *Journal of the Royal Society of Medicine* 58, no. 2 (1965): 85–96.
Roberts, Andy, *Albion Dreaming: A Popular History of LSD in Britain* (Singapore: Marshall Cavendish International [Asia], 2008).
Lobanovska, M. and G. Pilla, "Penicillin's Discovery and Antibiotic Resistance: Lessons for the Future?" *Yale Journal of Biological Medicine* 90, no. 1 (March 2017): 135–45.
Bud, Robert, *Penicillin: Triumph and Tragedy* (New York: Oxford University Press, 2007).

Chapter 35

Birstein, Vadim, *The Perversion of Knowledge: The True Story of Soviet Science* (New York: Perseus Books Group, 2013).
Carlson, Elof A., *Genes, Radiation, and Society: The Life and Work of H. J. Muller* (Ithaca, NY: Cornell University Press, 1981).
———, "Speaking Out about the Social Implications of Science: The Uneven Legacy of H. J. Muller," *Genetics* 187, no. 1 (January 2011): 1–7.
Foster, John Bellamy, *The Ecological Revolution: Making Peace with the Planet* (New York: Monthly Review Press, 2009).
Pearce, Mark S. et al., "Radiation Exposure from CT Scans in Childhood and Subsequent Risk of Leukaemia and Brain Tumours: A Retrospective Cohort Study," *Lancet* 380, no. 9840 (August 4, 2012): 499–505.
Russell, Bertrand, *Man's Peril, 1954–55*, vol. 28 of *The Collected Papers of Bertrand Russell*, ed. Andrew G. Bone (London: Routledge, 2003).

Chapter 36

Cori, Carl F. , "The Call of Science," *Annual Review of Biochemistry* 38 (July 1969): 1–21.
Gardner, A. L., "Gerty Cori, Biochemist, 1896–1957," *Women Life Scientists: Past, Present, and Future—Connecting Role Models to the Classroom Curriculum*, ed. Marsha L. Matyas and Haley-Ann E. Oliphant (Rockville, MD: American Physiological Society, 1997).
Larner, Joseph, "Gerty Theresa Cori (1896–1957)," *Biographical Memoirs of the National Academy of Sciences*, vol. 61 (Washington, DC: National Academies Press, 1992.
Smeltzer, Robert et al., *Extraordinary Women in Science and Medicine: Four Centuries of Achievement* (New York: Grolier Club, 2013).

Chapter 37

Cohn, Barbara A. et al., "DDT Exposure in Utero and Breast Cancer," *Journal of Endocrinology and Metabolism* 100, no. 8 (2015): 2865–72.

Davis, Kenneth S., "The Deadly Dust: The Unhappy History of DDT," *American Heritage* 22, no. 2 (1971).

Läuger, P., H. Martin and P. Mueller, "Über Konstitution und toxische Wirkung von natürlichen und neuen synthetischen insektentötenden Stoffen" [On constitution and toxic effects of natural and new synthetic insecticides], *Helvetica Chimica Acta*, 27 (1): 892–928, https://onlinelibrary.wiley.com/doi/abs/10.1002/hlca.194402701115.

Lovett, Richard A., "Oceans Release DDT from Decades Ago: Emissions of Controversial Pesticide Are Heading Northwards," *Nature* (January 7, 2010), https://www.nature.com/news/2010/100107/full/news.2919.4htm.

McGrayne, Sharon B., *Prometheans in the Lab: Chemistry and the Making of the Modern World* (New York: McGraw-Hill, 2001).

Seidensticker, John C. and Harry V. Reynolds III, "The Nesting, Reproductive Performance, and Chlorinated Hydrocarbon Residues in the Red-Tailed Hawk and Great Horned Owl in South-Central Montana," *Wilson Bulletin* 83, no. 4 (December 1971): 408–18.

Sledge, Daniel and George Mohler, "Eliminating Malaria in the American South: An Analysis of the Decline of Malaria in 1930s Alabama," *American Journal of Public Health* 103, no. 8 (August 2013): 1381–92.

Soto, Ana M. and Carlos Sonnenschein, "Environmental Causes of Cancer: Endocrine Disruptors as Carcinogens," *National Review of Endocrinology* 6, no. 7 (July 2010): 363–70.

Chapter 38

The Animal's Defender and Zoophilist, vols. 16–17 (London: National Anti-Vivisection Society [Great Britain]), 1897, original obtained from University of Michigan, digitized December 2, 2008, https://catalog.hathitrust.rog/Record/000522901.

Doby, T., "Cerebral Angiography and Egas Moniz," *American Journal of Roentgenology* 159 (August 1992): 364.

Freeman, Walter, James W. Watts, and Waco VA Medical Center, *Psychosurgery: In the Treatment of Mental Disorders and Intractable Pain* (Springfield, IL: Charles C. Thomas, 1950).

Freeman, Walter, James W. Watts, and Waco VA Medical Center, *Psychosurgery: In the Treatment of Mental Disorders and Intractable Pain* (Springfield, IL: Charles C. Thomas, 1950).

Hess, C. W., "Walter R. Hess (17.3.1881-12.8.1973), *Schweizer Archiv für Neurologie und Psychiatrie* 159, no. 4 (April 2008): 255–61.

Jansson, Bengt, "Controversial Psychosurgery Resulted in a Nobel Prize," Nobel Media AB 2014, http://www.nobelprize.org/nobel_prizes/medicine/laureates/1949/moniz-article.html.

Larson, Kate C., *Rosemary: The Hidden Kennedy Daughter* (New York: Houghton Mifflin Harcourt, 2015).

Laurance, Jeremy, "Ten Things that Drive Psychiatrists to Distraction," *Independent*, March 19, 2001.

Nijensohn, D. E., "Prefrontal Lobotomy on Evita Was Done for Behavior/ Personality Modification, Not Just for Pain Control," *Neurosurgical Focus* 39, no. 1 (2015): E12.

Rowland, Lewis P., "Walter Freeman's Psychosurgery and Biological Psychiatry: A Cautionary Tale," *Neurology Today* 5, no. 4 (April 2005): 70–72.

Vogt, Lucile, Thomas S. Reichlin, Christina Nathues, and Hanno Würbel, "Authorization of Animal Experiments Is Based on Confidence Rather than Evidence of Scientific Rigor," *PLOS* [Public Library of Science] *Biology* 14, no. 12 (December 2, 2016): e2000598.

Chapter 39

Ingle, Dwight, "Edward C. Kendall," *Biographical Memoirs of the National Academy of Sciences*, vol. 47 (Washington, DC: National Academies Press, 1975).

Lloyd, M., "Philip Showalter Hench, 1896–1965," *Rheumatology* 41, no. 5 (May 2002): 582–84.

Le Fanu, James, *The Rise and Fall of Modern Medicine* (New York: Carroll & Graf Publishers, 2000).

Index

About the Author

Moira Dolan, MD, is a graduate of the University of Illinois School of Medicine and has been a practicing physician for over thirty years. Dr. Dolan is a patient advocate and public speaker who educates patients on their rights and the need for a healthy skepticism of the medical profession. She is a contributor to the blog SmartMEDinfo and the author of three previous books, *No-Nonsense Guide to Antibiotics, Dangers, Benefits & Proper Use*; *No-Nonsense Guide to Cholesterol Medications, Informed Consent and Statin Drugs*; and *No-Nonsense Guide to Psychiatric Drugs, Including Mental Effects of Common Non-Psych Medications*. Dr. Dolan maintains a private medical practice in Austin, Texas.